JOURNAL OF ICT STANDARDIZATION

Volume 2, No. 2 (November 2014)

Special Issue on
ITU Kaleidoscope 2014: "Towards 5G"

Guest Editor:
Mr. Malcolm Johnson,
Director of the Telecommunication
Standardization Bureau (TSB),
International Telecommunication Union (ITU)

Guest Associate Editor:
Ms. Alessia Magliarditi,
Programme Coordinator, Policy and Technology
Watch Division (PTWD), TSB, ITU

JOURNAL OF ICT STANDARDIZATION

Chairperson: Ramjee Prasad, CTIF, Aalborg University, Denmark
Editor-in-Chief: Anand R. Prasad, NEC, Japan
Advisors: Bilel Jamoussi, ITU, Switzerland
Jesper Jerlang, Dansk Standard, Denmark

Editorial Board
Kiritkumar Lathia, Independent ICT Consultant, UK
Hermann Brandt, ETSI, France
Kohei Satoh, ARIB, Japan
Sunghyun Choi, Seoul National University, South Korea
Ashutosh Dutta, AT&T, USA
Alf Zugenmaier, University of Applied Sciences Munich, Germany
Julien Laganier, Luminate Wireless, Inc., USA
John Buford, Avaya, USA
Monique Morrow, Cisco, Switzerland
Vijay K. Gurbani, Alcatel Lucent, USA
Henk J. de Vries, Rotterdam School of Management,
Erasmus University, The Netherlands
Yoichi Maeda, TTC Japan
Debabrata Das, IIIT-Bangalore, India
Signe Annette Bøgh, Dansk Standard, Denmark
Rajarathnam Chandramouli, Stevens Institute of Technology, USA

Objectives

- Bring papers on new developments, innovations and standards to the readers
- Cover pre-development, including technologies with potential of becoming a standard, as well as developed / deployed standards
- Publish on-going work including work with potential of becoming a standard technology
- Publish papers giving explanation of standardization and innovation process and the link between standardization and innovation
- Publish tutorial type papers giving new comers a understanding of standardization and innovation

Aims

- The aim of this journal is to publish standardized as well as related work making "standards" accessible to a wide public – from practitioners to new comers.
- The journal aims at publishing in-depth as well as overview work including papers discussing standardization process and those helping new comers to understand how standards work.

Scope

- Bring up-to-date information regarding standardization in the field of Information and Communication Technology (ICT) covering all protocol layers and technologies in the field.

JOURNAL OF ICT STANDARDIZATION

Volume 2, No. 2 (November 2014)

Published, sold and distributed by:
River Publishers
Niels Jernes Vej 10
9220 Aalborg Ø
Denmark

www.riverpublishers.com

Journal of ICT Standardization is published three times a year. Publication programme, 2013–2014: Volume 2 (3 issues)

ISSN: 2245-800X (Print Version)
ISSN: 2246-0853 (Online Version)
ISBN: 978-87-93237-26-1

Editorial Foreword

This issue, 2nd issue of 2nd volume (2nd year), of the journal brings a selection of papers covering several hot-topics in standards and technology as listed below:

- This is the 1st issue in planned 4 special issues from ITU Kaleidoscope 2014.
- The issue also features a paper by Dr. Marcus Wong on the important topic of security assurance. This paper covers both technical challenges and on-going standardization activities.

The last issue of volume 2 to be published in March 2015 is the 2nd special issue of ITU Kaleidoscope 2014 and also a special issue on Network Function Virtualization (NFV) / Software Defined Networks (SDN).

We thank you for your interest in the journal and to the authors for their continuous support by submitting papers to the journal.

Editor-in-chief Journal for ICT Standardization
Anand R. Prasad, Dr. Ir., NEC Corporation, Japan

Guest Editor Foreword

The Kaleidoscope academic conference is ITU's flagship academic event. Established in 2008, the conference has been welcomed with great enthusiasm by the ICT research community and has matured into one of the highlights of the ITU's calendar of events.

Kaleidoscope's principal aim is to shed light on information and communication technology (ICT) research at an early stage so as to identify associated standardization needs in the interests of promoting the widespread diffusion of research findings through the development of internationally recognized ITU standards.

The sixth edition of the conference, ITU Kaleidoscope 2014, took the theme "Living in a converged world - impossible without standards?". It was tackled from a variety of perspectives in line with the multidisciplinary approach mandated by the all-encompassing nature of technological and industrial convergence.

From a total of ninety-eight submissions from thirty-nine countries, a double-blind, peer-review process selected thirty-four papers for presentation at the conference, all of which were published in the Kaleidoscope Proceedings as well as the IEEE *Xplore* Digital Library.

This is the first in a series of four special issues to showcase extended versions of selected Kaleidoscope papers, with the series planned as follows:

1. Special issue-1: Towards 5G
2. Special issue-2: e-Health and Standards
3. Special issue-3: Standardization, Education and Innovation
4. Special issue-4: Assessments, Models and Evaluation

This first special issue, "Towards 5G", includes a collection of Kaleidoscope papers that address questions essential to the development and rollout of '5G' mobile-wireless technologies.

Topics covered pertain to a range of subjects relevant to the development of the 5G ecosystem, spanning from services in cyberspace to the Internet of Things, dynamic use of radiofrequency spectrum, and security. The five papers and their respective authors are as follows:

- Proposal of "Cyber Parallel Traffic World" Cloud Service.

  Yoshitoshi Murata (Iwate Prefectural University, Japan); Shinya Saito (Iwate Prefectural University Graduate School, Japan)
- Standardizing the Internet of Things in an evolutionary way.

  Subin Shen (Nanjing University of Posts and Telecommunications, P.R. China); Marco Carugi (China Unicom, France)
- A Non-cooperative TV White Space Broadband Market Model for Rural Entrepreneurs.

  Sindiso Mpenyu Nleya (Computer Science Department, South Africa); Bigomokero Antoine Bagula (University of Western Cape, South Africa); Marco Zennaro (ICTP - The Abdus Salam International Centre for Theoretical Physics, Italy); Ermanno Pietrosemoli (International Centre for Theoretical Physics (ICTP), Italy)
- Performance evaluation of a dual diversity reception base on OFDM RoFSO systems over correlated log-normal fading channel.

  Fan Bai (Waseda University, Japan); Yuwei Su (Waseda University, Japan); Takuro Sato (Waseda University, Japan)
- A Mutual Key Agreement Protocol To Mitigate Replaying Attack In eXpressive Internet Architecture (XIA).

  Beny Nugraha (Mercu Buana University, Indonesia); Rahamatullah Khondoker (Fraunhofer SIT, Germany); Ronald Marx (Fraunhofer SIT, Germany); Kpatcha Bayarou (Fraunhofer Institute for Secure Information Technology, Germany)

We would like to thank the authors for their preparation of extended papers, the papers' reviewers for their generous contribution of time and expertise, and of course the readers of this journal for their interest and feedback on this first Kaleidoscope special issue.

Guest Editor: Mr. Malcolm Johnson
                      Director of the Telecommunication Standardization
                      Bureau (TSB), International Telecommunication
                      Union (ITU)

Guest Associate
Editor: Ms. Alessia Magliarditi,
                      Programme Coordinator, Policy and Technology
                      Watch Division (PTWD), TSB, ITU

"Cyber Parallel Traffic World" Cloud Service in 5G Mobile Networks

Yoshitoshi Murata[1] and Shinya Saito[2]

[1]Professor, Faculty of Software and Information Science, Iwate Prefectural University, Takizawa, Iwate, Japan, y-murata@iwate-pu.ac.jp
[2]Graduate School of Software and Information Science, Iwate Prefectural University Graduate School, Japan, g231j017@s.iwate-pu.ac.jp

Received: October 17, 2014;
Accepted: November 10, 2014

Abstract

Remarkable progress has been made in intelligent transportation systems. For example, autonomous vehicles can now detect their positions, as well as those of other vehicles, pedestrians, and obstacles. Pedestrians with smartphones can also determine their positions and send this information to the Internet. However, even though vehicles and pedestrians can instantly determine their positions, the information used for this purpose is rarely sent to a cloud service. Here, we propose a new cloud service, called "cyber parallel traffic world" (CPTW), in which vehicles, pedestrians, and temporary obstacles exist and move in synchronization with their real-world counterparts. In addition, virtual vehicles and pedestrians in the CPTW can communicate with real/virtual vehicles and pedestrians. The CPTW service could make driving safer and less stressful; provide people with experience in driving on roads throughout the world, etc. We describe how to construct its virtual world and establish communications.

Keywords: ITS, smart mobility, cloud service, safety drive, virtual world, 5G mobile networks, Internet of Things, IOT.

Journal of ICT, Vol. 2, 65–86.
doi: 10.13052/jicts2245-800X.221

1 Introduction

Remarkable progress has been made in intelligent transportation systems (ITSs). For example, autonomous vehicles can now detect their positions by using advanced global navigation satellite systems (GNSS). Moreover, they can detect other vehicles, pedestrians, obstacles, and traffic signals by using radar sensing, camera systems, and so on, and can drive themselves to the destinations specified by their drivers [1,2]. Several of these capabilities, such as obstacle detection, have already been implemented in commercial vehicles. Information on the positions of vehicles and pedestrians can be gathered from optical beacons in the universal traffic management system (UTMS) and from roadside "ITS Spots" (communication units) and can be used to prevent traffic accidents, predict traffic flows, control vehicle movements, and so on [3–5]. In Toyota's vision of a "smart mobility society," an inter-vehicle communication system provides every vehicle with communication links with the vehicles ahead and behind, enabling them to maintain a safe distance from each other. This will help to reduce traffic congestion and make travel safer and less stressful [6].

We proposed a cloud service, called "cyber parallel traffic world" (CPTW), for one of the future ITSs [7]. This service also would be one of the Internet of Things (IOT) [8,9] which work on the 5G mobile networks. In the CPTW roads, sidewalks, and traffic facilities such as traffic signals are represented as they are in the real world. Vehicles, pedestrians, and temporary obstacles exist and move in synchronization with their real-world counterparts. Virtual vehicles can also be driven in CPTW as if they were in the real world represented in the cloud. The second feature is different from what is directly possible in the real world: vehicles, drivers and pedestrians can communicate with other vehicles, drivers, and pedestrians by pointing to their positions rather than referring to an ID (e.g., a telephone number or address). For example, drivers can communicate with pedestrians, by giving messages such as "Go ahead and cross the street", and drive while talking with other drivers around them. Vehicles detect the positions of other vehicles and pedestrians, and avoid traffic congestion in CPTW. In addition, communications such as these enable the motions of the vehicles to be coordinated so as to reduce their emissions. This helps to make driving safer and less stressful.

We anticipate that a CPTW can be constructed using position information sent from the next generation mobile terminal, that is a successor of the smartphone, and vehicle navigation systems in the near future, and thus, it will not need information from proprietary traffic facilities such as UTMS optical beacons. This will make the CPTW system more cost effective than similar existing systems. The CPTW requires high-speed constant Internet connection, a flexible group communication and relay communication. However, even LTE, that is the latest mobile network, does not provide either group communications or relay communications; its characteristic of the fast hand off would not be sufficient for the high-speed constant Internet connection [10]. We hope that 5G mobile networks or other networks will be able to provide the high-speed constant Internet connection and these communication forms.

We have already developed a method for constructing virtual roads and traffic signals and defining traffic rules for them. Three-dimensional models of roads and traffic signals are made from elemental data such as road width measurements stored in a database system of a client's mobile terminal or vehicle navigation system.

After discussing related work in Section 2, we describe the concept and features of CPTW in Section 3. The construction of roads and the establishment of traffic rules are explained in Section 4. The communication modes and development scheme are described in Section 5 and 6. The key points of this paper are summarized in Section 7.

2 Related Work

Since our proposal is related to ITS and 3D road modelling, we will describe representative technologies, i.e., UTMS, which embodies the latest ITS technologies, and Forum8's 3D road modelling system, called VR-Drive.

2.1 UTMS

Since many of the CPTW functions are the same as those in UTMS, we will start with an overview of UTMS [3,4]. UTMS was developed by the UTMS Society of Japan, which has many Japanese manufacturers and organizations as members, to improve road traffic safety, smooth traffic flows, and ultimately harmonize road traffic with the environment so as to contribute to public welfare. UTMS provides nine main functions:

- The Integrated Traffic Control System (ITCS) uses technologies such as infrared beacons and computers to provide optimum signal control for effectively dealing with ever-changing traffic flow patterns and providing real-time traffic information.
- The Advanced Mobile Information System (AMIS) gathers real-time traffic information and provides it to drivers.

- The Public Transportation Priority System (PTPS) gives priority to public transportation by monitoring bus lanes, warning vehicles that are illegally using the bus lane, and pre-empting traffic signals.
- The Mobile Operation Control System (MOCS) helps transportation administrators manage their bus, freight, sanitation, and other operations.
- The Environment Protection Management System (EPMS) reduces traffic pollution (exhaust gas and noise) and thus helps protect the regional environment.
- The Driving Safety Support System (DSSS) helps drivers drive more safely by providing them with information gathered using various sensors that detect vehicles and pedestrians.
- The "Help system for Emergency Life saving and Public safety" (HELP) immediately reports position information to rescue organizations in the case of an emergency such as a traffic accident, vehicle breakdown, or sudden illness.
- The Pedestrian Information and Communication System (PICS) helps the elderly and the disabled to move around safely.
- The Fast Emergency Vehicle Pre-emption System (FAST) helps emergency vehicles reach an accident site as quickly as possible.

There are many functions in UTMS for improving traffic safety, smoothing traffic flow, and protecting the environment. However, the information used in it is mainly gathered using proprietary intelligent traffic facilities. Therefore, establishing a UTMS is expensive.

Autonomous vehicles can determine not only their own positions but also the positions of other vehicles, pedestrians, and obstacles. Pedestrians who have smartphones can also determine their positions. If this position information could be collected and presented in a simulated world, i.e., CPTW, on the Internet and if an inter-vehicle communication function were implemented, many of the UTMS functions described above could be provided at much lower cost.

2.2 Forum8's VR-Drive

The 3D modelling environment that we want should be able to extract road data, traffic rules data, etc., from real-world sources, put these data into road and traffic rules database, and create 3D roads, etc., by referring to these databases. The trouble is that most modelling tools can't extract the sort of data we need from real-world sources.

An exception is VR-Drive [11] developed by Forum8. This tool simulates real-world driving environments in every possible way, and includes, e.g., the traffic volume, pedestrians, cyclists, junctions, traffic lights, hills, holes, rail crossings, ITS road signs, and roundabouts. It is developed for addressing the following issues:

- Developing safe and eco-friendly driving schemes.
- Highlighting lapses of attention to improve drivers' awareness.
- Training drivers in special driving procedures.
- Examining the behaviour, specifically, the attention of drivers.
- Preparing drivers to face and anticipate any driving situation (e.g., accident scene, slippery road surface, etc.).

Forum8 provides road databases to create 3D road environments for Japan and several other countries.

Since the VR-Drive has many functions, it is one of best pieces of 3D road-environment software that could be chosen to complete CPTW. However, its license fee is expensive and its software is not open source. Therefore, we had to develop a new 3D road modelling software by ourselves.

3 Concepts and Features of the CPTW

Roads, traffic facilities, temporary obstacles, and sidewalks in the real world are represented as 2D or 3D models in CPTW as shown in Figure 1. There are also traffic rules, which are the same as in the real world, and it is possible to detect whether the corresponding real/virtual vehicles and pedestrians obey them. Data on the positions of pedestrians and vehicles detected by the next generation mobile terminals and vehicle navigation systems are sent to CPTW, where they are plotted in real time. Position data that other vehicles and intelligent traffic facilities detect are also plotted in real time. Virtual vehicles are also presented in a way that distinguishes them from the corresponding real-world vehicles. This feature would be useful, say, for people practicing driving in another country where they have had no experience driving. Moreover, a real/virtual driver and pedestrian can communicate with other drivers and pedestrians by pointing to a position on the vehicle navigation or the next generation mobile terminal screen, not an ID (e.g., phone number or address). This scheme is unlike existing communication schemes and has not been developed yet. People could construct a local CPTW for their place of residence. These local worlds could in turn be connected to create a larger and eventually global CPTW.

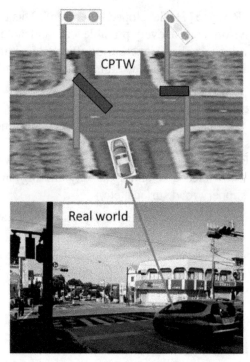

Figure 1   Images showing correspondence between CPTW and the real world

Such a global CPTW could assist drivers and pedestrians in several ways.

- They could be made aware of vehicles and pedestrians screened by obstacles.
- A real/virtual driver could see the environment of the road and roadside through a real vehicle's video camera.
- A real driver could send another real driver or pedestrian a message such as "Go ahead" or "Be careful a little way ahead." Moreover, real/virtual drivers could drive a vehicle while talking with other real/virtual drivers. This would enable drivers to be more courteous.
- It is very difficult for a driver to keep his or her car in a lane during snowfalls or heavy rain at night. However, once the positional accuracy gathered in real-time can be made finer than a few tens of cm, a driver will be able to know how well his or her vehicle is keeping within a lane in such conditions.
- A virtual driver could gain experience driving on roads throughout the world by using CPTW as a driving simulator. This would be particularly

useful to people planning to visit and drive in another country where they have had no experience driving.

- A real driver could be warned that she or he is approaching the scene of a traffic accident, enabling her or him to take appropriate action and avoid secondary accidents.
- Real vehicles could drive closer to one another, which would cut wind resistance and reduce fuel consumption.

4 Constructing the CPTW

We have to construct a virtual world encompassing all the roads, traffic facilities, and sidewalks throughout the real world. Real/virtual vehicles and pedestrians who are residents in the CPTW have to be plotted in this virtual world.

In this virtual world, it should be possible to detect whether a real/virtual vehicle or pedestrian is obeying the traffic rules of the corresponding country. These virtual structures can be constructed by extracting road data from maps covering the world, such as Google Maps and Google Earth, and by gathering the traffic rules for each country. These data would then be stored in a roads database and a rules database. An application program would create 2D/3D virtual roads by using data in the roads database and place traffic signals and signs in accordance with the rules in the traffic rules database, as shown in Figure 2 [12].

We used OpenGL as the 3D program interface and developed a program using the "glut," "sdl", "glew," and "OpenAL" tools [13–15].

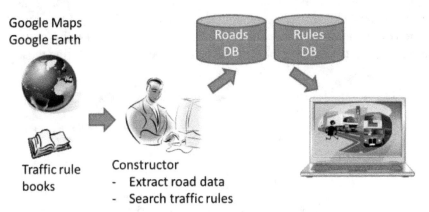

Figure 2 Overview of constructing CPTW

4.1 Roads

A road is represented as an approximate function for detecting whether a vehicle or pedestrian is obeying the traffic rules.

First, the simulation program finds the direction perpendicular to the parametric curve that expresses the centre line of the road in order to create the width of an approximated road. It then calculates the coordinates of a point shifted to the right or left of the centre line, as shown in Figure 3.

The tangential angle at an arbitrary point on the curve can be calculated as

$$\theta = Tan^{-1}\frac{dy}{dx}. \tag{1}$$

A point at the road edge is one that is shifted to the road edge from an arbitrary point (x, y) on the curve. Points (x_r, y_r) at the right side road edge and points (x_l, y_l) at the left side road edge can be calculated as

$$x_r(t) = W_r \cos(\theta - \frac{\pi}{2}) + x(t) \tag{2}$$

$$y_r(t) = W_r \sin(\theta - \frac{\pi}{2}) + y(t) \tag{3}$$

$$x_l(t) = W_l \cos(\theta + \frac{\pi}{2}) + x(t) \tag{4}$$

$$y_l(t) = W_l \sin(\theta + \frac{\pi}{2}) + y(t) \tag{5}$$

W_r is the width of the right side on the road and W_l is the one on the left. We use the 3D spline curve as an approximate function for roads except intersections. Since the 3D spline curves pass through the plotted points, as shown in Figure 4, it is easy to extract road data. Each part of 3D road polygons is created by changing the parameter "t" of the three-dimensional spline curve from zero to one, calculating many points on the road edge, and storing those

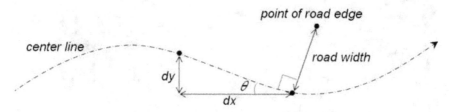

Figure 3   Method for creating 3D roads

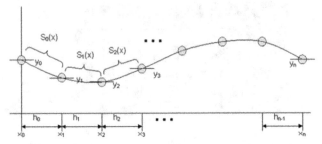

Figure 4 Three-dimensional spline curves on plotted points

points in a vertex array. Road centre lines and lane lines are created by changing the value of W in Equations (1) – (5).

When the width or number of lanes at $t = 0$ differs from that at $t = 1$, the simulation program decides that the road has a right- or left-turn-only lane. When a road has such a lane, the road width is gradually increased. The simulation program does this by calculating a smooth curve that represents increments in the lane width. A sigmoid function is used to increase the width. It is a monotonically increasing function and has one inflection point. Therefore, it is suitable for expressing a right- or left-turn-only lane. An example of a 3D straight road with an increasing number of lanes is shown in Figure 5 (a).

Two-dimensional B-spline curves are used as functions for intersections. The polygon for an intersection consists of all curve functions that connect to the intersection. The parameter "t" of the calculated curve function is changed from zero to one in the same way as for a road, and an intersection polygon is created. An example of an intersection is shown in Figure 5 (b).

At least, following data have to be stored in the roads database.
[General roads]

- Road ID: identifies road
- 3rd coefficient of curve X
- 2nd coefficient of curve X
- 1st coefficient of curve X
- 0th coefficient of curve X
- 3rd coefficient of curve Y
- 2nd coefficient of curve Y
- 1st coefficient of curve Y
- 0th coefficient of curve Y
- Right-side width at $t = 0$
- Left-side width at $t = 0$

(a) Straight road

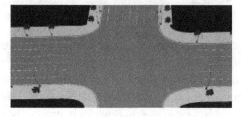

(b) Intersection

Figure 5   Examples of created 3D roads

- Right-side width at t = 1
- Left-side width at t = 1
- Number of right-side lanes at t = 0
- Number of left-side lanes at t = 0
- Number of right-side lanes at t = 1
- Number of left-side lanes at t = 1
- ID of intersection at t = 0
- ID of intersection at t = 1

[Intersections]

- Intersection ID
- Curve ID
- 2nd coefficient of intersection curve X
- 1st coefficient of intersection curve X
- 0th coefficient of intersection curve X
- 2nd coefficient of intersection curve Y
- 1st coefficient of intersection curve Y
- 0th coefficient of intersection curve Y

Sidewalks are constructed by adding thickness to the road edge parameter in Equations (2) – (5). Roadside feature and traffic facilities other than signals, such as trees and lights lining a street, are created using 3D modelling tools and placed on sidewalks. Their positions are calculated using Equations (2) – (5)

4.2 Traffic Rules

The traffic rules are classified as either "unspecified rules," such "keep to the right," or "specified rules" derived from traffic signals and signs. The specified rules are stored in the rules database. The CPTW detects whether vehicles and pedestrians at an intersection or on a road under the control of a traffic signal and/or sign are obeying the rules. For example, a vehicle at an intersection where a "no turn on red" sign (Figure 6) is posted is checked to see whether it turns right when the controlling traffic signal is red.

A scheme is used to relate traffic signals to the meanings of signs. The scheme has to handle traffic signals and rules worldwide.

We collected traffic rules and traffic signal definitions for three countries: Japan, the U.S., and Germany. The colors of signal lights and their basic meanings have been defined by the CIE (Commission Internationale de l'Éclairage) and the ISO (International Organization for Standardization) [16,17]. Green means "go," yellow means "prepare to stop," and red means "stop." These colors and meanings are common worldwide.

There are, however, several local rules in each country. For example, in Japan, when a green arrow light attached to a green/yellow/red light signal switches on, a vehicle is permitted to turn in the direction of the arrow even

Figure 6   Traffic sign for "No Turn on Red"

if the signal is red or yellow. A yellow arrow prohibits a vehicle from not only going in the direction of the arrow but also straight. However, a tram can proceed in the direction of the arrow. A red flashing light permits a vehicle to proceed after first stopping and checking for other vehicles. A yellow flashing light permits a vehicle to proceed if doing so does not interfere with the movement of other traffic.

A study of the traffic rules in the U.S. (based on the driver handbooks for California, Florida, and New York [18–20]) reveals that a vehicle is usually allowed to turn right even if the controlling signal is red unless there is a posted traffic sign prohibiting such a turn (Figure 6). The meanings of flashing red and yellow are the same as in Japan. In addition to the green arrow, there are red and yellow arrows. The meaning of the green arrow light is the same as in Japan. A red arrow prohibits going straight but permits turning right or left. A yellow arrow indicates that the green arrow will soon end, so vehicles should prepare to stop. Japan's green arrow has a similar meaning.

A traffic signal in Germany consists of a green, yellow, and red light [21], the same as in Japan and the U.S. Whereas the red light changes directly to green in the U.S. and Japan, in Germany it changes to green after the red and yellow are lit simultaneously. A steady red light prohibits a vehicle from proceeding, the same as in Japan. However, if a green arrow traffic sign is attached to the traffic signal, as shown in Figure 7, turning right is permitted even if the red light is on. When the green arrow light attached to a regular

Figure 7   Traffic signs attached to traffic signal (Germany)

traffic signal switches on, turning in the direction of the arrow is permitted. This is the same as with Japan and USA's green arrow light. When the yellow arrow light attached to a regular traffic signal switches on, turning in the direction of the arrow is permitted if doing so does not interfere with the movement of other traffic. This is different form USA and Japan's yellow arrow light. If the traffic signal has only a red and yellow light, both of them being switched off corresponds to a steady green light. A flashing yellow light permits a driver to proceed after first stopping and checking for other vehicles. These signals are a little different from the rules in Japan and the USA.

Some traffic signals are attached with the once stopping sign as shown in Figure 8. If there is a contradiction between a traffic signal and a traffic sign, a traffic signal is superior to a traffic sign.

There are two basic types of light used in traffic signals worldwide: round and arrow-shaped. The colors are green, yellow, and red. The conditions for the round type are "on," "off," and "flashing." The various combinations of light type, light color, condition, and traffic sign have different meanings that may also differ between countries. Despite the variety, the meanings of these combinations are generally related to permission or prohibition of a vehicle proceeding or to taking caution.

Table 1 describes the four-layer traffic rules model, and Figure 8 illustrates how we can ascertain a real traffic rule from a real object. Each combination of elements is assigned a traffic rule. For example, the combination of a red light and a green arrow is assigned a rule stating that a vehicle is permitted to turn right as a general rule. For the case shown in Figure 9, there is a "No Turn on Red" sign at the intersection. We assign this "No Turn on Red" sign a rule stating that a vehicle is prohibited from turning right as a general rule. We create a real traffic rule by combining these two general rules and considering the priority of each general rule.

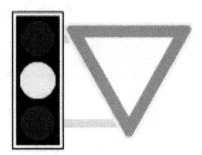

Figure 8   Traffic sign attached to traffic signal (Germany)

Table 1   Traffic rules model

Layer Name	Function
Real rule layer	Create real rules by combining general rules
General rule layer	Assign traffic rules to various combinations of elements
Significant object layer	Combine objects that have significant rules
Real object layer	Decompose traffic signals or traffic signs into elements (objects) such as light type, color, and condition

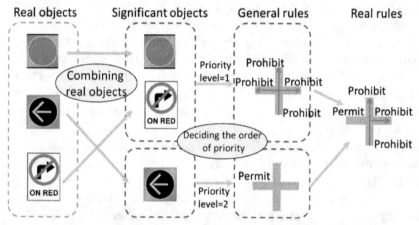

Figure 9   Illustration of defining real traffic rules from real objects

4.3 Plotting Objects in the Virtual World

Remarkable progress has also been made in GNSSs. In addition to the American Global Positioning System, we now have Russia's GLONASS, the EU's Galileo, and Japan's quasi-zenith satellite system, "Michibiki." In the near future, measurement accuracies of less than 1 m will be available to general users [22–24]. The next generation mobile terminal and vehicle navigation system will then be able to determine their position within 1 m.

Every object logged in a CPTW keeps a session with the CPTW cloud servers and sends them positioning data. In addition, vehicles and pedestrians who drive and walk in a CPTW should have realistic aliases and avatars. For instance, an avatar should look like its user for ease of recognition.

Vehicles sometimes drive so fast that positioning data periodically sent from them would yield plotted positions different from their real ones. Here, we propose that a vehicle sends, in addition to its position, time stamps and three-dimensional data on its acceleration, and velocity. This information would enable the next generation mobile terminal or navigation system to estimate the relative positions of nearby vehicles in real-time, even though measurement accuracy is more than 1 m. We would re-plot every real vehicle in a way that it maintains its driving lane. In addition, the vehicle would detect the shapes of obstacles, the distance between them and itself, and send the detected shapes and distances to the CPTW servers.

Every traffic signal in the CPTW has to be synchronized with the real ones. A set-top box on a traffic signal would be a good way to detect a signal change. However, this method is impossible at present. Another idea is to store its change periods, and synchronize them with the mobility of vehicles and pedestrians near the traffic signal. Further study is needed to determine whether this idea is useful or not.

5 Communication Schemes

As we mentioned in the introduction, the CPTW needs the following communication modes:

1. One-to-one mode: a real/virtual vehicle, driver or pedestrian communicates with another real/virtual vehicle, driver or pedestrian by pointing to their avatars or saying its relative position or nearby feature. This communication mode would be used for general communications. In particular, the CPTW system could use it to gather information such as additional positioning data or fuel consumption data from the vehicle navigation system and/or next generation mobile terminal.
2. Group mode: multiple real/virtual drivers and pedestrians talk to each other. The originating object does not change, but people are free to join or leave a group. This communication mode would be used for talking with other real/virtual drivers and pedestrians.
3. Relay mode: a real vehicle exchanges information with vehicles ahead of and behind it. This communication mode would be used for tandem driving to cut wind resistance and reduce fuel consumption. Virtual vehicles would not be included in such a line of vehicles.

The CPTW requires the high-speed constant Internet connection characteristics to plot vehicles or pedestrians precisely and the last two functions of the CPTW written in Section 3 also requires very low-latency and high reliability transmission characteristics.

Conventional cellular mobile networks do not as yet have above group mode or relay mode (they have the one-to-one mode). And their transmission characteristics are not sufficient for warning or avoiding traffic accidents. Trunked radio systems support group communications [25]. Unfortunately, the members of each group in a trunked radio system are fixed and cannot be changed freely. In addition, virtual drivers and pedestrians are not distinguished from real drivers in a communication group. Meanwhile, neither cellular mobile networks nor trunked radio systems support a relay mode. However, there are many studies on incorporating mobile networks in ITSs and for inter-vehicle communications in the real world [26]. ITU collaborates with organizations related to ITS to standardize its communications [27]. Some of their studies and standards are related to the above group mode and relay mode.

We propose the virtual communication modes shown in Figure 10. We will assume that every object in the CPTW keeps a session with the CPTW server as in Skype. In one–to-one mode, a real/virtual driver points to an avatar on a screen or says the features or relative position of a destination. Then he or she selects talking or messaging, and both the sessions of the driver and the

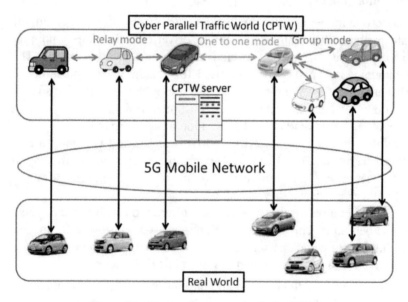

Figure 10 Communication modes in the CPTW

indicated avatar or object are connected through the CPTW system. Avatars engaged in a conversation would change some aspect of their appearance, such as their color. When the other real/virtual drivers and pedestrians point to a central avatar, their sessions would all be connected and a communication group would be formed; this is the group mode. In the case of relay mode, a driver selects a form of usage such as tandem driving, and control signals are relayed within the line of real vehicles through the CPTW system. However, the positions of the vehicles cannot always be guaranteed to be the same as the real ones. Thus, their group mode and relay mode has difficulty in warning about the traffic accident just happened to the next vehicle coming if it's already within 20 to 30 meters from the accident site. For this reason, we hope that the 5G mobile networks of the future will shorten the transmission latency and provide the flexible group communication mode and relay mode.

6 Development Scheme

Our aim is to provide a standardized CPTW for use by people worldwide, in both developed and developing countries, at little or no cost and enable them to create a CPTW representing their area of residence. Therefore, the system cost should be as low as possible. This means that expensive traffic facilities such as UTMS ones must be excluded and that most of the information needed must be gathered through the next generation mobile terminals and vehicle-mounted devices cooperating with residents in the world. CPTW is targeted for resident participation design, meaning that local residents construct the parts of CPTW in which they live. This will reduce the cost and make CPTW more realistic.

Since many issues including problems caused by communication networks described in Section 5 need to be resolved, it is very difficult for us to develop CPTW by ourselves.

We need many people and the ITU-T to help us. The best approach is thus to use open sources and an open standard scheme.

The issues to be addressed include

1. deciding the specifications that include services;

   - Service content,
   - Protocol between client devices and the CPTW servers,
   - Protocol between in-vehicle devices and others,
   - Synchronization scheme between vertical traffic signals and real traffic signals, -etc.,

2. deciding how to verify the developed technologies, software, and equipment to ensure they meet the specifications, and
3. deciding how to provide the service and manage its operation.

7 Conclusion

In the proposed cyber parallel traffic world (CPTW) cloud service, roads, sidewalks, and traffic facilities such as traffic signals are presented as they are in the real world. Vehicles, pedestrians, and temporary obstacles move in synchronization with their real-world movements. Since the next generation mobile terminals will soon be able to determine their positions to accuracy within 1 m and most vehicles will be able to detect not only their own position but also the positions of other vehicles and obstacles, it should be possible to gather most of the information needed through the next generation mobile terminals and vehicle-mounted devices. The virtual structures would be constructed by extracting road data from maps and by gathering traffic rules for each country, storing the data in a roads database and a rules database, and using an application program to create 2D/3D roads using data in the roads database. This will make it possible to provide many useful and attractive services at a reasonable cost. In this paper, we also discussed how to realize the one-to-one communication mode, the group communication mode, and the relay communication mode of the CPTW system. Although business models for the CPTW were not discussed in this paper, since the CTPW connects the real world and the virtual world, it suits the O2O (Online to Offline) service well, and we can expect the creation of new business models.

References

[1] Erico Guizzo, "How Google's Self-Driving Car Works," IEEE Spectrum, February 26, 2013.
[2] Autonomos Labs, http://www.autonomos.inf.fu-berlin.de/, October 2014.
[3] ITS GREEN SAFETY SHOWCASE, http://www.its-jp.org/english/its-green-safety-showcase/, October 2014.
[4] UTMS Society of Japan, http://www.utms.or.jp/english/index.html, October 2014.
[5] T. Oda, K. Takeuchi, Y. Yoshio, S. Niikura, "Evaluation of measured travel time utilizing two-way communication in UTMS," IEEE, VNIS ' 96 (Volume: 7), pp. 54 – 61, Oct. 1996.

[6] TOYOTA SMART MOBILITY SOCIETY, http://www.toyota-global .com/innovation/smart_mobility_society/, October 2014.

[7] Yoshitoshi Murata, Shinya Saito, "Proposal of "Cyber Parallel Traffic World" Cloud Service", ITU Kaleidoscope 2014, S1.4. 2014.

[8] Chonggang Wang, etc., "Internet of Things," IEEE, 2014

[9] Internet of Things Global Standards Initiative, http://www.itu.int/en/ITU-T/gsi/iot/Pages/default.aspx

[10] G. Araniti, C. Campolo, M. Condoluci, A. Iera, and A. Molinaro, "LTE for vehicular networking: a survey", IEEE Communications Magazine, Vol. 51, No. 5, pp. 148–157, May 2013.

[11] Forum8, engineering 3D environment, http://www.forum8.com/default .asp, October 2014.

[12] Shinya Saito, Yoshitoshi Murata, Tsuyoshi Takayama and Nobuyoshi Sato, "An International Driving Simulator - Recognizing the Sense of a Car Body by the Simulator -," Workshops in AINA 2012, W-FINA-S12.1, pp. 254–260, March 2012.

[13] OpenGL – The Industry Standard for High Performance Graphics, http://www.opengl.org/, October 2014.

[14] Simple DirectMedia Layer, http://www.libsdl.org/, October 2014.

[15] Home – OpenAL, http://openal-soft.org/, October 2014.

[16] CIE, CIE S 006.1/E-1998, "Road Traffic Lights - Photometric Properties of 200 mm Roundel Signals", http://www.cie.co.at/Publications/index. php?i_ca_id=467, October 2014.

[17] ISO, ISO 16508:1999," Road traffic lights – Photometric properties of 200 mm roundel signals", http://www.iso.org/iso/home/store/catalogue _tc/catalogue_detail.htm?csnumber=31003, October 2014.

[18] California Driver Handbook - Traffic Lights and Signs, http://apps.dmv.ca. gov/pubs/hdbk/traff_lgts_sgns.htm, October 2014.

[19] Florida Drivers Handbook - Traffic Signals, http://www.123driving.com/ flhandbook/flhb-traffic-signals.shtml, October 2014.

[20] Driver Handbook in New York, http://www.dmv.org/ny-new-york/driver- handbook.php#Drivers-Manual, October 2014.

[21] How to Germany – Driving in Germany -, http://www.howtogermany.com /pages/driving.html, October 2014.

[22] Groves, Paul D, "Principles of GNSS, Inertial, and Multisensor Integrated Navigation Systems: Second Edition," ISBN=9781608070053, Artech House Publishers, 2013 John

[23] M. Dow, R. E. Neilan, C. Rizos, "The International GNSS Service in a changing landscape of Global Navigation Satellite Systems," Journal of Geodesy, Volume 83, Issue 3–4, pp. 191–198, March 2009.
[24] Y Li, C Rizos, "Evaluation of positioning accuracy of GNSS with QZSS augmentation," IGNSS Symposium 2011, November 2011.
[25] Trunked Radio Communication System, http://wiki.radioreference.com/index.php/Trunking, October 2014.
[26] Mihail L. Sichitiu, Maria Kihl, "Inter-Vehicle Communication System: A Survey," IEEE Communication Surveys & Tutorials, 2nd Quarter, 2008.
[27] Collaboration on ITS Communication Standard, http://www.itu.int/en/ITU-T/extcoop/cits/Pages/default.aspx

Biographies

Yoshitoshi Murata received his M.E from Nagoya University, Japan. He received his Ph.D from Shizuoka University, Japan. From 1979 to 2006, he was belonging to NTT and NTT DoCoMo and developing mobile communication systems, mobile terminals and services. Since 2006, he is a professor of faculty of Software and Information Science, Iwate Prefectural University. Prof. Murata was awarded the best paper of the ITU-T "Innovations in NGN" Kaleidoscope Academic Conference 2008. His research interests include mobile communication, sensor usage application and ITS. He is a member of IEEE, IEICE and IPSJ.

Shinya Saito received his M.E. from Iwate Prefectural University. Since 2013, he is belonging to Information System Dept. in Tokyo Electron Limited and involved in the work of ERP system and Business Intelligence.

An Evolutionary Way to Standardize the Internet of Things

Subin Shen[1] and Marco Carugi[2]

[1]*School of Computer, Nanjing University of Posts and Telecommunications, Nanjing, China*
[2]*Study Group 13, International Telecommunication Union, Telecommunication Standardization Section, Geneva, Switzerland*
sbshen@njupt.edu.cn, marco.carugi@gmail.com

Received: October 17, 2014;
Accepted: November 10, 2014

Abstract

The current situation of technology separation among the different application domains of the Internet of Things (IoT) results in a market separation per application domain. This issue hinders the technical innovation and investments in the IoT business. In order to solve the issue, it is necessary to standardize common technologies of the IoT across the different application domains. This paper argues that a key direction of the future standardization of the IoT, in addition to standardizing specific technologies, is building over a standardized new architecture reference model for the IoT. Based on the analysis of existing key activities concerning the standardization of OSI, NGN and IoT from a functional architecture perspective, it suggests that the IoT standardization work be progressed in an evolutionary way in order to enable the integration of existing technologies, and focus on the interactions among functional entities or deployable components of the IoT to impose minimum constraints on future technical innovations. Topics discussed in the paper include characteristic capabilities of the IoT, ways of integrating the cloud computing technologies in the IoT perspective, and challenges faced by the IoT standardization work.

Journal of ICT, Vol. 2, 87–108.
doi: 10.13052/jicts2245-800X.222

Keywords: architecture reference model, functional entity, interaction, Internet of Things, Next Generation Network, Open System Interconnection, standardization.

1 Background

The International Telecommunication Union (ITU) had identified in its Internet Reports 2005 on the Internet of Things (IoT) that "a new dimension has been added to the world of information and communication technologies (ICTs): from *anytime*, *anyplace* connectivity for *anyone*," to "connectivity for *anything*" [1]. The evolution from anytime and anyplace to anything connectivity is a development goal of ICTs in the near future.

Now, there seems to be a long way to reach this goal. Even though the IoT technologies have been applied in various application domains, the IoT has not been developed in the view of a global information infrastructure. There is no standardized IoT architecture for global support of the different application domains. From the industrial point of view, the technologies of the IoT are segmented by the different application domains, so that the market of the IoT is also divided by the different application domains. This situation hinders the investment and development of IoT technologies and applications. In this sense, significant advances of the IoT standardization are essential and urgent for the development of IoT technologies and applications.

It has to be admitted that the IoT is indeed a very complex subject of the information and communication technologies. And the standardization of the IoT is then a very complex work too. There are different perspectives and opinions about the IoT standardization work: some focus on the specific technologies of the IoT in order to make the IoT of practical value, such as the work on machine-to-machine communications (M2M) done by the European Telecommunications Standards Institute (ETSI, www.etsi.org); others focus on a standardization process starting with the IoT requirements and capabilities and moving then to the IoT architecture, such as the work on IoT architecture reference model done by the IoT-A (Internet of Things – Architecture) project (www.iot-a.eu) co-funded by the European Commission within the Seventh Framework Programme (2007–2013).

In our opinion, in addition to the standardization of the specific technologies of the IoT, it is necessary to standardize a new architecture reference model (ARM) for the IoT, as the existing reference models are not suitable for the IoT. This paper assumes that the Open System Interconnection (OSI) basic reference model (BRM) [2], the Next Generation Network (NGN) BRM [3] and the NGN functional architecture [4] are the key existing reference models over which the new ARM for the IoT can be developed.

The standardized IoT ARM should enable the integration of existing technologies and support the unique requirements of the IoT applications.

In order to enable the integration of existing technologies, it is suggested that the standardized IoT ARM builds over the OSI BRM, the NGN BRM and the NGN functional architecture, and the standardization work focuses on the interactions among the IoT functional entities or deployable components in order to impose minimum constraints on future technical innovations.

In order to support the unique requirements of the IoT applications, the characteristic requirements and corresponding characteristics capabilities of the IoT should be analyzed at first. Then, the ARM of the IoT can be standardized building on existing reference models and the characteristic capabilities of the IoT.

The contents of this paper are arranged as follows.

- In Section 2, the evolution of the standardization requirements in an architecture perspective is presented based on the analysis of the inter-actions in the OSI BRM, NGN BRM and NGN functional architecture, and the interactions which are required to be standardized in the IoT are then described.
- In Section 3, some characteristic capabilities of the IoT are introduced.
- Section 4 presents the evolution of the ARM based on the analysis of the frameworks of OSI BRM, NGN BRM, and NGN functional architecture, and a proposed framework of the IoT ARM for IoT standardization is then discussed.
- In Section 5, ways of integrating the cloud computing technologies in the IoT perspective are discussed.
- Section 6 discusses some existing activities (outside the ITU-T) related to the standardization of the IoT.
- In Section 7, challenges faced by the IoT standardization work are analyzed.
- In Section 8, some suggestions for the future standardization of the IoT are given.
- The conclusion of this paper is given in Section 9.

2 The Evolution of the Standardization Requirements

The analysis of the requirements for the standardization of the architecture reference model is the first step in the standardization activity. The standardization requirements had evolved from those of the OSI BRM to those of the NGN BRM from the perspective of the interactions.

The OSI BRM aims to *"qualify standards for the exchange of information among systems"* [2]. The exchange of information can be termed as interactions, and the systems can be regarded as the computing systems. So the OSI BRM is required to guide the standardization of the interactions among computing systems. The interactions in the OSI BRM are illustrated in Figure 1, where the standardized interactions are represented by solid arrows and other interactions are represented by dashed arrows. The application process is *"an element within a real open system which performs the information processing for a particular application"*.

The OSI BRM is a logical reference model that does not consider the implementation and deployment aspects (i.e. specific service access points for each functional layer, the functional components in each functional layer, etc.), so it is not suitable to guide directly the standardization of the technologies for implementation and deployment.

The NGN BRM takes into account the requirements concerning the adoption of existing transport and service technologies and the separation between service system (service functions) and transport system (transport functions). So, as illustrated in Figure 2, the NGN functional architecture is required to standardize the interactions among multiple service systems, among multiple transport systems and between service system and transport system of the same NGN. It should be noted that, in Figure 2, an application is *"a structured set of capabilities, which provide value-added functionality supported by one or more services"* [5].

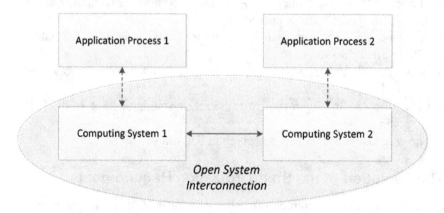

Figure 1   Interactions in the OSI BRM

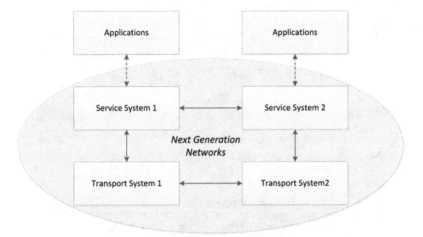

Figure 2   Interactions in the NGN BRM

The separation of service system from transport system results in independent service provisioning in NGN, so that a competitive environment is promoted in order to speed up the provisioning of diversified NGN services [4]. In this sense, the standardization requirements have evolved from the information exchange in OSI BRM to the service provisioning in NGN BRM.

Concerning the IoT, the interactions are extended to the interactions between computing system and IoT devices; in these interactions the exchanged information is constrained to the information collected by the devices and the information controlling the devices. Here the IoT devices refer to the devices connected to the IoT. As the messages exchanged between computing system and IoT devices are no longer transparent in terms of information meaning, it is required to identify and process the meaning of the information exchanged from the application perspective. These content-aware interactions between computing system and IoT devices can be classified into local interactions and remote interactions.

The local interactions are illustrated in Figure 3. This type of interactions may be initiated by the local computing system or the local related applications. As far as the technologies related with this type of interactions, existing technologies can be used, such as those developed for wireless sensor networks. This is the basic interaction type to be standardized in the IoT.

The remote interactions are illustrated in Figure 4. This type of interaction is decomposed into the local interactions between IoT devices and local

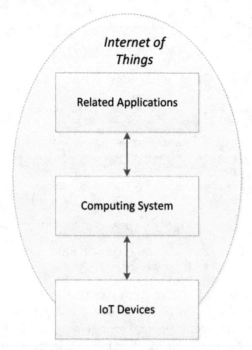

Figure 3   Local interactions in the IoT

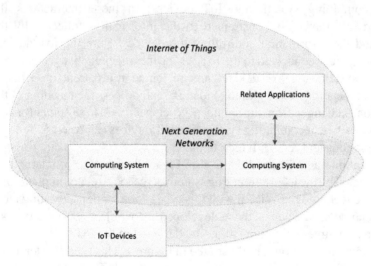

Figure 4   Remote interactions in the IoT

computing system, and the interactions between local computing system and remote computing system as described in the NGN specifications and their necessary extensions to be specified in IoT related standards. NOTE - It is assumed here that the IoT is implemented building over the NGN.

This type of interactions is a complex part of the IoT standardization, requireing to integrate IoT technical standards and NGN technical standards.

Based on the above analysis, a global view of the interactions in the IoT is proposed as illustrated in Figure 5, where the computing system is decomposed into transport system and service system as defined in the NGN specifications. The standardization of the different interactions shown in the figure (between IoT devices and transport system, between transport system and service system, between service system and related applications, among different transport systems and among different service systems) is required.

Although the IoT devices in Figure 5 are illustrated as directly connected to the transport system, the IoT devices need to interact with the service system and even related applications based on the requirements of IoT applications. An IoT device can be also regarded as a specific computing system that is required to automatically interact with the transport system, the service system and related applications, possibly resulting in automatic service provisioning.

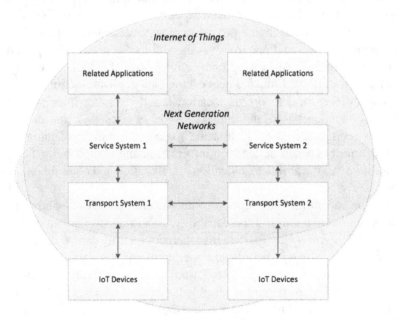

Figure 5   A global view of the interactions in the IoT

Also, the IoT devices exchange data with the related applications, and these interactions belong to a content-aware type of information exchange.

In IoT, the standardization requirements have then evolved from a content-transparent information exchange as assumed in the OSI BRM, and the traditional service provisioning as assumed in the NGN BRM, to a content-aware information exchange and the automatic service provisioning as required by the IoT applications. In this sense, the IoT requires a new set of characteristic capabilities and a new standardized architecture reference model to fulfill its characteristic requirements.

3 Characteristic Capabilities of the IoT

According to the characteristic requirements of the IoT, the standardization work of the IoT involves a new set of characteristic capabilities.

Building on some characteristic requirements identified in ITU-T studies [10,11], significant examples of these characteristic capabilities of the IoT are, in our opinion, the capabilities for support of autonomic operations, location based operations and data operations.

The capabilities for support of autonomic operations include the capability for autonomic service provision and that one for autonomic networking. The first one involves the abilities of automatic capture, transfer and analysis of data of things, as well as the automatic provision of services based on predefined rules or policies. The second one involves the abilities of networking parameter self-configuration, self-healing connectivity with the IoT, networking performance self-optimization, and networking entity self-protection.

The capabilities for support of location based operations include the capability for location based and context aware services, and that one for location based communication. The first one involves the abilities of automatic provision of services based on location information and related context. The second one involves the abilities of location identification and communication initiation.

Among the capabilities for support of data operations, the capability for semantic based data operations, the capability for virtual storage and virtual processing, and the capability for data management security can be identified. The capability for semantic based data operations involves the abilities of semantic based annotation, query, storage, transfer, and aggregation of data of things in order to fulfill the requirements of IoT applications. The capability for virtual storage and virtual processing involves the abilities of providing storage and processing resources in a dynamic and scalable way. The capability for data management security involves the abilities of providing secure, trusted and privacy protected data management.

4 The Evolution of the ARM

The ARM of the IoT can be seen as an evolution of existing reference models. In this paper we consider the OSI BRM, the NGN BRM and the NGN functional architecture as the key existing reference models.

The OSI BRM is independent from implementations and deployments, and was standardized from a logical perspective. The OSI basic reference model consists of four basic elements: "open systems" that are pertinent to OSI, "application-entities" that exist within the OSI environment, "associations" that connect the application-entities and permit them to exchange information, and "physical media for OSI" that provide the means for the transfer of information between open systems [2].

The NGN BRM was adopted in the NGN functional architecture [4]. The NGN functional architecture is defined as "*a set of functional entities and the reference points between them used to describe the structure of an NGN. These functional entities are separated by reference points, and thus, they define the distribution of functions*" [4]. Although the reference points are sufficient to guide the standardization of the local interactions between service system and transport system, it is our opinion that functional entities and reference points are not sufficient for satisfying the requirements of remote interactions, such as the remote service provisioning requirements. The reason of such limitation is that the architecture of a system consists of its functional entities and the interactions among them, and a reference point refers to "*a conceptual point at the conjunction of two non-overlapping functional entities that can be used to identify the type of information passing between these functional entities*" [4]. Functional entities and reference points cannot cover all interactions among the various functional entities, such as the interactions among functional entities connected through networks (in fact, for these functional entities, the associations - in OSI BRM's terms - among these functional entities should be also considered besides the reference points).

Although each functional entity can be specified by its capabilities and its reference points with other functional entities, a functional architecture would require specification of the functional entities, the reference points and the associations among the reference points.

The NGN functional architecture specified in [4] has evolved from the OSI BRM by taking into account requirements and capabilities from the implementation and deployment perspectives, with separation of the service functions from the transport functions. The OSI BRM is however still used in the modeling of the NGN as a supporting model of communication.

The NGN functional architecture includes NGN functional entities and NGN components [4]. The definition of functional entity in [4] is *"an entity that comprises an indivisible set of specific functions. Functional entities are logical concepts, while groupings of functional entities are used to describe practical, physical implementations."* Although a definition of component (NGN component) is not given in [4], based on the content of clause 10 of [4] and the above definition of functional entity, the components can be regarded as the "groupings of functional entities which are used to describe practical, physical implementations." In this sense, "component" is a concept that is used in deployment models.

From the above analysis, it can be concluded that [4] contains both the specifications of functional model of NGN and deployment model of NGN. The functional model of NGN is specified from a logical point of view, and the deployment model of NGN is specified from a deployment point of view.

Figure 6 illustrates the framework of the "NGN ARM" consisting of OSI BRM as communication model, NGN functional model and NGN deployment model.

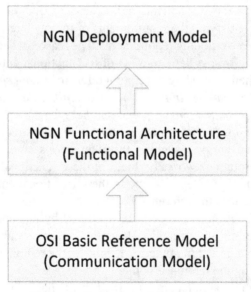

Figure 6   Framework of the NGN ARM

As the standardization requirements of the IoT include the standardization of content-aware interactions and automatic service provisioning, the only supporting model of communication is not sufficient for the ARM of the IoT. Three additional models are identified for the standardization of the ARM of the IoT: the information model, the security & privacy model, and the IoT concept and requirement model. A framework of the IoT ARM is proposed and is illustrated in Figure 7.

The technical scope of the IoT covers different aspects of ICTs, such as computing, control and communication technologies. In order to define correctly and reasonably the scope of the IoT standardization, the establishment of an IoT concept and requirement model is necessary to specify characteristics and high level requirements of the IoT from an ICT perspective.

Based on the requirements of IoT applications, the IoT is required to automatically capture and process (e.g., classify, aggregate, store, transfer and analyze) the data collected by the IoT devices, and possibly send instructions for their control according to pre-defined rules and occurrences of pre-defined events. The IoT depends on the understanding of the meaning of the captured or transferred data for correctly performing the required operations. In order to standardize the information exchange among IoT functional entities from the application perspective and support automatic operations in the IoT,

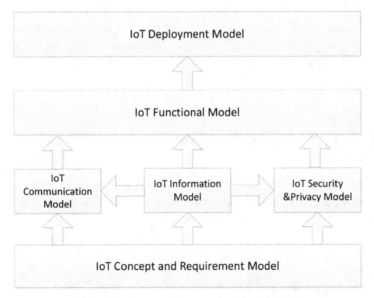

Figure 7   Proposed framework of the IoT ARM

it is necessary to establish an IoT information model specifying the IoT information entities and their relations based on the application requirements. An information model has been proposed by the IoT-A project. However, as the IoT information model involves some complex topics, such as the data models related with physical things and the semantics of physical things, more work is expected to be done in the future.

The security and privacy protection capabilities are no longer optional in the IoT. A security & privacy model is needed in order to specify the constraints of security and privacy protection distributed in the IoT information entities, and the operations to enforce and fulfill these constraints. "Concepts and solutions for privacy and security in the resolution infrastructure" of the IoT have been proposed by the IoT-A project, but these concepts and solutions are not linked to the IoT information model.

In our opinion, in order to integrate the different models of the IoT ARM, a "knowledge plane" such as that one proposed by David Clark in [6] may be needed in the IoT functional model. This means that the functional model should have the capabilities of capturing, storing, using and updating the knowledge in support of the operations required in the IoT. This knowledge could be represented in the IoT functional model according to formal descriptions contained in the information model.

In summary, the ARM has evolved from the logical model specified in the OSI BRM, and the logical and deployment oriented model specified in the NGN functional architecture, to the conceptual, logical and deployment oriented model required for the IoT. Because this paper focuses on the trend for standardizing the IoT and the standardization work concerning the IoT ARM is at an early stage, the paper does not go further into details of the IoT ARM.

5 Integration of Cloud Computing Technologies with the IoT

Based on the framework of the IoT ARM, ways of integrating other (IoT relevant) technologies in the IoT perspective can be described clearly. The integration of existing technologies with the IoT impacts in fact the content of one or more of the models identified for the standardization of the IoT ARM.

In this paper, we take the cloud computing technologies as an example of existing technologies that can be integrated with the IoT in order to support data operations and flexible service provisioning.

As mentioned earlier, the capability for virtual storage is required in the IoT to enable a dynamic and efficient usage by the IoT functional entities of all possible storage resources, within the IoT or connected to the IoT, for storage of the data of things. Similarly, the capability for virtual processing is required in the IoT to enable a dynamic and efficient usage by the IoT functional components of all possible processing resources, within the IoT or connected to the IoT, for processing the data of things.

These two capabilities of the IoT can be implemented based on cloud computing technologies. These capabilities can be also supported in the IoT service platform that can then integrate its services with cloud computing oriented services.

The capability for support of "Things as a Service" is an example of additional capabilities which can be provided in the IoT via the integration of cloud computing technologies with the IoT. This capability involves the abilities of publishing interfaces of virtual things as services, and linking virtual things with physical things in order to implement functionalities of sensing and actuating things by the invocation of corresponding services.

6 Some Existing Activities (Outside the ITU-T) Concerning the Standardization of the IoT

Several relevant efforts concerning the standardization of the IoT have been progressed and/or are in progress besides that one conducted within the Internet of Things Global Standard Initiative (IoT-GSI) administrated by the ITU-T (www.itu.int/en/ITU-T/gsi/iot). This paper simply concentrates on two of these efforts, the "Smart Machine-to-Machine communications (SmartM2M)" ETSI (www.etsi.org) Technical Committee (continuation of the previous "M2M" Technical Committee) and the IoT-A project (www.iot-a.eu).

ETSI has published technical specifications concerning the M2M functional architecture, respectively in 2011 and 2013.

The ETSI "M2M service capabilities functional architecture framework" published in 2011, illustrated in Figure 8, provides a network-independent architecture framework for M2M [7]. It adopts the principle of separation between service functions and transport functions as specified in the NGN functional architecture [4], and focuses on the service layer aspects.

The ETSI "M2M service capabilities functional architecture framework" published in 2013, illustrated in Figure 9, extends the previous framework in some deployment aspects [8].

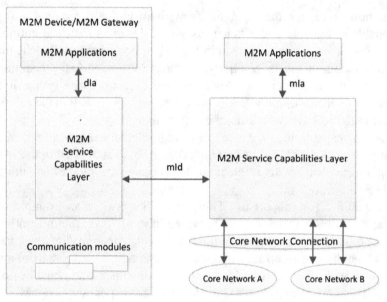

Figure 8   ETSI M2M service capabilities functional architecture framework published in 2011

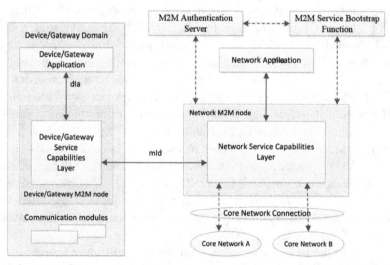

Figure 9   ETSI M2M service capabilities functional architecture framework published in 2013

It can be realized that the ETSI architecture framework published in 2013 has introduced the concept of node ("Device/Gateway M2M node" and "Network M2M node") in order to emphasize the deployment aspects of the architecture framework, although a clear distinction is not made there between functional aspects and deployment aspects. As already anticipated, functional model and deployment model serve different purposes, and their specifications should be addressed in a distinct way.

The IoT-A project has done a lot of studies on the IoT ARM. The IoT-A ideas and methodology concerning the study of the IoT ARM are - in our opinion - very valuable for further research and standardization activities concerning the IoT ARM. The IoT ARM has been decomposed into IoT domain model, IoT information model, IoT functional model, IoT communication model, and IoT trust, security & privacy model. The framework of the IoT ARM defined by the IoT-A project [9]is illustrated in Figure 10.

In this framework, differently from the proposed framework of the IoT ARM shown in Figure 7, either the communication model or the trust, security & privacy model do not support the IoT functional model, they are simply included within the IoT functional model. This approach may be satisfactory from a research point of view, but may have some limitations from a technical standardization point of view.

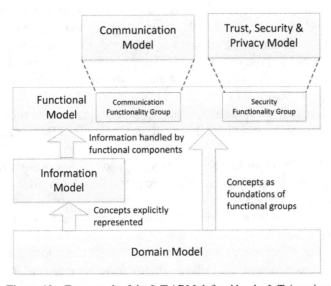

Figure 10   Framework of the IoT ARM defined by the IoT-A project

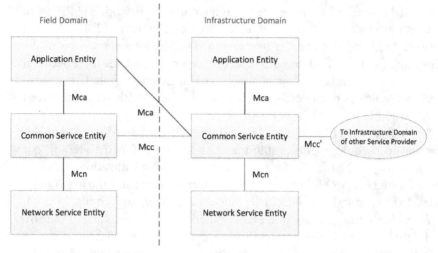

Figure 11   oneM2M functional architecture

As a relatively new effort concerning the standardization of the IoT, it is worthwhile to mention here also oneM2M, a global standards partnership for Machine to Machine Communications and the Internet of Things. As reported in the official website of the oneM2M organization (www.onem2m.org), "the purpose and goal of oneM2M is to develop technical specifications which address the need for a common M2M Service Layer that can be readily embedded within various hardware and software, and relied upon to connect the myriad of devices in the field with M2M application servers worldwide." In summary, oneM2M focuses on the standardization of a common M2M Service Layer that constitutes a part of the IoT architecture.

Among its initial developments, oneM2M is progressing a functional architecture specification: Figure 11 illustrates the oneM2M Functional Architecture [12].

As far as the entities in Figure 11 are concerned: an Application Entity represents an instantiation of Application logic for end-to-end M2M solutions; a Common Services Entity represents an instantiation of a set of "common service functions" of the M2M environments; a Network Services Entity provides services from the underlying network to the Common Service Entities.

As far as the reference points in Figure 11 are concerned, they represent the reference points for M2M communications between relevant entities within a M2M Service Provider or among different M2M Service Providers.

7 Challenges in IoT Standardization

Many IoT technologies are still under development, and a large scale adoption of a given technology usually depends also on the availability of related standards. More globally, the availability of IoT standards of general applicability will be key to the large scale adoption of IoT solutions.

The IoT standardization work faces numerous challenges and this paper only highlights few significant challenging areas: autonomic capabilities, data operations and privacy protection. The availability of IoT standards of general applicability in these three areas involves the development of a standardized ARM of the IoT, as well as the standardization of specific technologies of the IoT.

The autonomic capabilities include capabilities not only for automatic operations, but also for other operations such as self-configuring, self-healing, self-optimizing and self-protecting operations. The implementation of these capabilities depends on the application of the autonomic computing theory, which is currently still a research topic. A challenge faced in the standardization of autonomic capabilities is then how to specify reasonable standards without constraining the technical innovations on autonomic capabilities in the future.

Data operations include not only the data operations for the IoT management and control, but also the data operations for IoT application support and service provision. A challenge faced in the standardization of data operations is how to specify standards for data operations which are independent of the different application domains.

In order to protect the IoT from harmful attacks, it is expected that every IoT user and every IoT device are somehow identified. How to marry identification requirements with proper privacy protection, and consequently how to specify adequate standards, is a real challenge in the standardization of the IoT.

8 Suggestions About the Future Standardization of the IoT

Based on the above analysis and discussions, some principles for the standardization of the IoT have been derived [13]. The following suggestions are given for consideration in the future standardization of the IoT in addition to the standardization of specific technologies.

Suggestion 1: The standardization of the IoT should focus on the interactions among functional entities from a functional perspective and among deployable components from a deployment perspective. This would simplify the standardization work and impose minimum constraints on future technical innovations.

According to the concept of standardization of the IoT based on the interactions among computing systems (assuming here that a related application can also be regarded as a computing system, so the interactions among computing systems include the interactions between related applications and computing system), and the interactions between computing system and IoT devices, the standardization of the IoT should focus on the capabilities related to the interactions. Some capabilities that are not related to the interactions, such as, for example, the capability for internal storage, should not be the focus of the IoT standardization work.

As an example of this suggestion to focus on interaction aspects, let's consider the capabilities for support of semantics in the IoT. If the related standardization work would focus on interaction aspects such as the exchange formats of semantically described data and the service provision using semantics, this would reduce the standardization work complexity and still leave space for future technical innovations via the research studies on semantics.

Suggestion 2: The standardization of the IoT should also focus on the standardization of the IoT ARM, taking into consideration the various models (i.e., according to the proposed framework of the IoT ARM described earlier, IoT concept and requirement model, IoT information model, IoT communication model, IoT security & privacy model, IoT functional model and IoT deployment model).

It is suggested that the standardization of the IoT ARM be specified at both logical and physical level. The physical level refers here to the deployment oriented level, and the logical lever refers to the deployment independent level. If possible, it is recommended that the logical level specifications be separated from the physical level specifications in order to simplify the technical specifications and provide stability and long term applicability of the logical level specifications.

In order to guide correctly the future IoT standardization work, it is suggested that the standardized IoT ARM be verified in its correctness and completeness. Verifying the IoT ARM is a complex task, but it is worth doing, including for the promotion of technical innovations. In this sense, it is very critical to attract both industry experts and academic experts in the standardization effort concerning the IoT ARM.

Based on this principle of correctness and completeness verification, with specific reference to the work carried out by the ITU-T IoT-GSI, the IoT standardization work on the IoT ARM should build over the specifications concerning IoT requirements [10,11] and IoT functional framework and capabilities (study currently in progress).

Suggestion 3: The standardization of the IoT should also focus on appropriate work concerning the deployment of the IoT in various application domains (e.g. e-health, intelligent transport systems, smart energy, smart cities), and the integration of IoT relevant technologies (e.g. cloud computing, software defined networking, web services, service delivery platform, data storage, data mining and other big data technologies), in order to strengthen the direct applicability and value of the IoT standards in the industry.

9 Conclusion

Based on the analysis of existing key activities concerning the standardization of OSI, NGN and IoT from a functional architecture perspective, a global view of the interactions in the IoT and a framework of the IoT ARM have been proposed. Also, an evolutionary way to define standard architecture reference models from OSI and NGN to IoT has been analyzed.

Based on these proposals and analysis, it is suggested that, in addition to the standardization of specific technologies, the future standardization of the IoT focuses on the architecture reference model of the IoT, as a guidance for the whole standardization process, and on the interactions among functional entities or deployable components of the IoT, in order to simplify the standardization work and impose minimum constraints on future technical innovations. It is also suggested to progress the IoT standardization work concerning the deployment of the IoT in specific application domains as well as the integration of IoT relevant technologies, in order to strengthen applicability and value of the IoT standards.

Considering that the IoT builds on existing information and communication systems, the IoT standardization work should be conducted in an evolutionary way in order to enable the integration, at the greatest possible extent, of existing information and communication technologies.

References

[1] International Telecommunication Union (ITU), "ITU Internet Reports 2005: The Internet of Things", 2005.

[2] ITU-T Recommendation X.200 |ISO/IEC 7498-1:1994, "Information technology – Open Systems Interconnection – Basic Reference Model: The basic model", 1994.

[3] ITU-T Recommendation Y.2011, "General principles and general reference model for Next Generation Networks", 2004.

[4] ITU-T Recommendation Y.2012, "Functional requirements and architecture of the NGN", 2010.

[5] ITU-T Recommendation Y.101, "Global Information Infrastructure terminology: Terms and definitions", 2000.

[6] D. Clark, C. Partridge, J. Ramming and J. Wroclawski. "A knowledge plane for the Internet". Proceedings of ACM SIGCOMM2003, August 2003, pp. 3-10, 2003.

[7] ETSI TS 102 690 v1.1.1, "Machine-to-Machine communications (M2M); Functional architecture", 2011.

[8] ETSI TS 102 690 v1.2.1, "Machine-to-Machine communications (M2M); Functional architecture", 2013.

[9] IoT-A Deliverable D1.4, "Converged architectural reference model for the IoT v2.0", 2012. http://www.iot-a.eu/public/public-documents/documents-1 (visited on 2014-02-26)

[10] ITU-T Recommendation Y.2060, "Overview of the Internet of Things", 2012.

[11] ITU-T Recommendation Y.2066, "Common requirements of the Internet of Things", 2014.

[12] oneM2M, oneM2M-TS-0001 - V-2014-08, "oneM2M Functional Architecture Baseline Draft", 2014-08-01.

[13] S. Shen, M. Carugi, "Standardizing the Internet of Things in an Evolutionary Way", Kaleidoscope 2014.

Biographies

Subin SHEN received the bachelor's, master's and PhD degrees in computer science and engineering from the Southeast University (SEU), Nanjing, China.

He is a professor in the School of Computer and the School of Software at the Nanjing University of Posts and Telecommunications (NUPT), China. His research interests include computer networks, telecommunication networks, the Internet of Things, cloud computing, big data, and future networks. He is one of editors of Recommendation Y.2066 "Common requirements of Internet of Things", Draft Recommendation "IoT functional framework and capabilities", and Draft Recommendation "IoT application support models" of International Telecommunication Union, Telecommunication Standardization Sector (ITU-T). He is a member of the Institute of Electrical and Electronics Engineers (IEEE), Association for Computing Machinery (ACM), China Computer Federation (CCF), and China Institute of Communications (CIC).

Marco Carugi Rapporteur for Question 2 of Study Group 13, International Telecommunication Union, Telecommunication Standardization Section, Geneva, Switzerland

Marco Carugi is currently Independent Consultant on advanced telecommunication technologies and associated standardization. During his professional career, he has worked as Telecommunication Engineer in the Solvay group, as Research Engineer in Orange Labs, as Senior Advisor in the Nortel Networks CTO division and as Senior Expert in the Technology Strategy department of ZTE R&D.

He is active in standardization since 1997, leading the development of numerous standards specifications and holding numerous leadership positions, including ITU-T SG13 vice-chair, Rapporteur for ITU-T Questions and ad-hoc groups, OIF Board member, IETF Provider Provisioned VPN working group co-chair. Currently, he is Rapporteur for Question 2 - "Requirements for NGN evolution and its capabilities including support of IoT and SDN" - inside ITU-T SG13 (Future networks), acts as SG13 Mentor and leads the development of technical specifications on requirements, capabilities and services for IoT/M2M in the ITU-T Internet of Things Global Standards Initiative.

Future Networks, SDP, SDN and Cloud Computing are other technical areas in which he is involved at present.

Marco holds an Electronic Engineering degree in Telecommunications from the University of Pisa in Italy, a M.S. in Engineering and Management of Telecommunication Networks from the National Institute of Telecommunications (INT) in France and a Master in International Business Development from the ESSEC Business School in Paris.

Optimisation of a TV White Space Broadband Market Model for Rural Entrepreneurs

Sindiso M Nleya[1], Antoine Bagula[2], Marco Zennaro[3]
and Ermmano Pietrosemoli[3]

[1]*University of Cape Town, ISAT Lab, South Africa, snleya@cs.uct.ac.za,*
[2]*University of Western Cape, ISAT Lab, South Africa, bbagula@uwc.ac.za*
[3]*The Abdus Salam International Centre for Theoretical Physics mzenarro, Italy,*
epietros@ictp.it

Received: October 16, 2014;
Accepted: November 10, 2014

Abstract

Leveraging on recent TV white space communications developments in regulations, standards initiatives and technology, this paper considers a suitable next generation network comprising of two primary users (PUs) that compete to offer a service to a group of secondary users (SUs) in the form of mesh routers that belong to different entrepreneurs participating in a non-cooperative TV white space trading. From a game theoretic perspective the non-cooperative interaction of the PUs is viewed as a pricing problem wherein each PU strives to maximize its own profit. Subsequently the problem is formulated as a Bertrand game in an oligopolistic market where the PUs are players who are responsible for selling TV white spectrum in the market while the SUs are the players who are the buyers of the TV white spectrum. The PUs strategise by way of price adjustment, so much such that SUs tend to favour the lowest price when buying. The inter- operator agreements are based on the delay and throughput QoS performance optimization metrics respectively. A performance evaluation of both models is comparatively performed with regards to parameters such as cost, generated revenue, profit, best response in

Journal of ICT, Vol. 2, 109–128.
doi: 10.13052/jicts2245-800X.223

price adjustments and channel quality. The throughput based analytic model fares better in terms of providing channel quality as it has a better strategy which is a decreased price value.

Keywords: White Spaces, Smart Radio, Non-Cooperative, Optimization, Game Theory, Broadband Market, Traffic Engineering.

1 Introduction

Ideally, in a free enterprise economy, entrepreneurs are compelled to provide telecommunication services when it is profitable. Certainly in poor areas, the prospects of profits are very minimal due to the poverty of potential customers such that there is little if any service. Intuitively there is a quest to identify sustainable means for closing the gap between service cost and the ability of customers to pay in areas with acceptable political and economic stability. Increasingly entrepreneurs and entrepreneurship are assuming a transformative role in the rural telecommunications economy and have the potential to narrow the gap between service cost and ability of customers to pay. Furthermore recent developmental trends in wireless technologies are not only providing various opportunities for entrepreneurs, but also overhauling the character of entrepreneurship by pioneering new business models. To date, an array of competing wireless technologies have entered the market and these range from Wireless Mesh technology, WiFi, WiMAX (802.16), Cellular such as UMTS/W-CDMA and High speed Downlink Packet Acess (HSPDA) [4] LTE and Advanced LTE. To this end, among these developments in the market, wireless mesh networks (WMNs), have indisputably and justifiably been touted as a candidate technology that is set to facilitate ubiquitous connectivity to the end user in underprivileged, underprovisioned, and remote areas. The WMNs comprise wireless routers and clients as well as an endowed ability to dynamically self organize, and self configure to the extent of nodes in the network being able to establish and maintain connectivity among themselves. The candidature of this technology justifiably emanates from its characteristic low upfront cost, ease of maintenance, robustness as well as reliable service coverage. Indisputably, WMNs have found applications ranging from broadband home networking, community and neighbourhood networks, enterprise networking, building automation and other public safety areas etc. However, while the currently deployed WMNs provide flexible and convenient services to the clients, the performance, growth and spread of

WMNs is still constrained by several design limitations [2] such as limited usable frequency resource. The design constraints are a consequence of WMNs in the unlicensed Industrial, Scientific and Medical (ISM) band being mostly adopted for access communications. Subsequently this adoption renders the WMN susceptible to competition with all other devices in this particular ISM band eg. nearby WLANS and Bluetooth devices. Ultimately, the limited bandwidth of the unlicensed bands cannot cope with the evolving network applications and this has led to the spectrum scarcity problem. To mitigate the impending spectrum scarcity problem, tangible efforts have been made to deregulate wireless spectrum resources and promote dynamic spectrum access (DSA). The regulatory aspect has evidently been the recent steps by the Federal Communications Comission (FCC) in opening up the Television (TV) spectrum band thereby allowing unlicensed devices to opportunistically access it as long as the unlicensed users do not interfere with legacy communications. The regulatory reform has been motivated partly by a series of empirical occupancy measurements that have revealed a gross under utilization of licensed spectrum, called white space, while on the other hand, the analog to digital Television transition has made available large chunks of spectrum called TV White Space (TVWS). To this end, the urge to exploit white spaces is irresistible as it provides an opportunity to significantly enhance the performance of WMNs and other wireless technnologies. Pursuant to the notion of harnessing TVWS, Smart Radio (SM) a device that has the capability to sense the environment and automatically adjust the configuration parameters has been proposed as a viable solution to the frequency reuse problem. Furthermore a fundamental application of SM is that of Dynamic Spectrum Access (DSA), a technique which allows SM radio to operate in the best available channel. Specifically, the SM radio technology will enable the users to [3]:

- Determine which portions of the spectrum is available and detect the presence of licensed users when a user operates in a licensed band.
- Select the best available channel (spectrum management).
- Coordinate access to this channel with other users (spectrum sharing) .
- Vacate the channel when a licensed user is detected (spectrum mobility).

A second constraint to the spread and growth of WMNs has been a case of many rural areas being still not deemed economically viable by operators. Service providers claim this is a result of dispersed populations, cost of roll-out and lack of power infrastructure which remains a hindrance to the efforts of service providers [6]. To this end, a wireless mesh network is thus equipped with Smart Radio devices gives rise to a Smart radio Mesh network

(SMWMN) which leverages on Dynamic spectrum acess. Ultimately dynamic spectrum access (DSA) wireless technology enables rural broadband internet service providers to access lower- frequency spectrum, reducing the cost of network deployment and operation. This will translate to service providers, for the first time being able to implement profitable business models and will provide consumers and businesses in rural areas with affordable and sustainable service [5]. According to [7], a combined decrease in the cost and increasing pervasiveness of access will have a positive social and economic impact in rural and remote areas. Moreover with SWMN holding the key to the last mile, the challenge is that of catalysing both decreased costs and increased access. An approach to this challenge involves leveraging on the common knowledge that telecommunications networks profit from network effects. The bigger the market the higher value it holds giving the incumbent (primary user) telecoms operator a massive strategic advantage. Essentially, the limited spectrum availed to mobile services translates to a constrained number of competitors in the market. Consequently, in many areas the effect has been a stagnation of competition and undesirably high telecommunications costs. Thus, increasing spectrum availability, in particular to new entrants, is likely to lead to more competition and healthier markets [5–6]. In this paper, we concentrate our efforts on modelling the competition in the rural telecommunication market in which the spectrum sharing technique is utilized within the context of a low cost Smart Wireless Mesh Network (SMWMN) for the provision of broadband internet services. More specifically, we extend our efforts in [9], to a non cooperative scenario in which network nodes belonging to different licensed wireless providers (PUs) engage in spectrum trade while competing to offer services to a secondary service and simultaneously striving to maximize profits. Furthermore we consider the Quality of Experience (QoE), which is a concept that is becoming widespread in the emerging network paradigms. Thus our contribution is as follows:

- We develop an analytic model for the design of a SWMN from a game theoretic perspective. Our SMWMN is formulated as a Bertrand duopoly market in which two PUs from varied wireless service providers compete with each other with regards to their prices so as to offer services to a secondary service. In the process the PUs are aiming to maximize their profits under quality of service (QoS) constraints.
- Adapt the model [15] to TV white space.
- Optimize the cost of sharing spectrum as a function of QoS degradation with the throughput as QoS performance measure.

- Comparative evaluation of the models in terms of the profit, cost, revenue, price strategy and channel quality.
- Predict the Quality of Experience (QOE) from a QoS perspective (delay and throughput).

The rest of the paper is organized as follows. Section II presents the related work which subsequently leads to a TV white space market pricing model in section III. Performance optimization of the models is presented in section IV and the conclusion as well as further work in section V.

2 Related Work

From a competitive market perspective, Niyato et al. [1] acknowledged the important role pricing plays in the trading of any resource or service. Basically the objective of trading is to provide benefits both to sellers and buyers. Thus the choice of a price must be motivated by the desire to simultaneously maximize revenue for the sellers (service providers) and satisfaction for the buyers (users). Pricing rules should be developed over open platforms that guarantee not only interoperability among the service providers, which would facilitate their cooperation, but also the implementation of their individual business strategies [10]. The choice of a price is influenced by the user demand and competition among service providers .

Within the context of Cognitive radio networks, pricing of spectrum resources has been addressed in numerous works [11–13]. In [11], a framework to facilitate dynamic spectrum access by way of an optimization problem approach formulated for the purpose of maximizing the revenue for the spectrum provider through pricing and spectrum assignment is presented. A scheme for competitive spectrum sharing wherein multiple self interested spectrum providers operating with different technologies and costs compete for potential customers is presented in [12] as a non cooperative game. A stochastic learning algorithm is implored to determine the Nash equilibrium which is itself a solution to this game. However, the authors did not consider the dynamics of a multi-hop cognitive wireless mesh network as well as the issue of resource allocation in this kind of network. However efforts involving multi-hop networks concentrate on spectrum sharing with interference aware transmission mechanism for each relay mechanism.

In [14], a Media Access Control (MAC) layer scheduling algorithm was proposed for a multi-hop wireless network. An integer linear programming model was formulated to obtain the optimal schedule in terms of time slot and channel to be accessed by the cognitive radio nodes. The problem of spectrum pricing and competition among primary users (or primary services) and interactions among the cognitive radios in a multi-hop mesh network were not considered in this work. Initiatives to focus on competitive spectrum sharing and pricing in cognitive wireless networks are recorded in [15]. The initiative involves two levels of competition, the first being among primary users and the second among secondary users for spectrum usage to choose the source rate to maximize their utilities. Non-cooperative games are formulated for these competitions with the Nash equilibrium being considered as the solution. Clearly, these efforts are not enough and can still be extended. Fang et al. [16] affirm that in addition to networking technologies, additional factors that determine the success of wireless mesh networks is whether there exists viable business models. There is limited research on this problem. In wireless mesh networks, wireless nodes are required to forward traffic for both themselves and their neighbours. If the nodes are controlled by self-interested users, they may not efficiently share their capacity to route traffic for other nodes. Such possibility undermines the performance and feasibility of wireless mesh networks, therefore effective pricing mechanisms need to be developed before mesh technologies are commercialized.

3 TV White Space Market Pricing Model

A. System Model

We present a competitive scenario within the context of spectrum management wherein licensed users of spectrum called primary users compete to offer services to an unlicensed users called secondary users. From a primary user perspective, the cost of providing a service to a secondary service is modeled as a function of Qos degradation. This being a game, Nash equilibrium is considered to be the optimal solution.

Bertrand model generally depicts competition for an oligopoly market scenario comprising a homogeneous product with static and non discriminatory prices. In the classical case, this model fits well for a scenario of two firms bidding in a project in which the winner subsequently takes the entire project. Alternatively two firms may attempt to dominate a market and each one of the firms has sufficient manufacturing capacity to make all the product. Ultimately the lowest bidder gets the business. We however adapt the model to deal with the spectrum market scenarios within the context of a SMWMN as shown in Figure 1. To begin with, a summary of the notation to be used in the ensuing analysis is presented in Table 1.

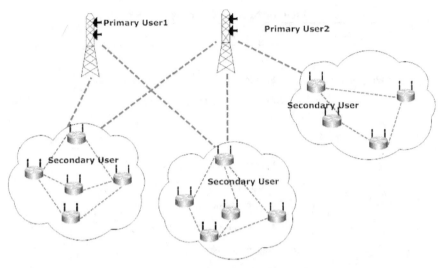

Figure 1 Smart Mesh Network

We consider the existence of N primary users operating on dissimilar frequency spectrum and a grouping of secondary users desiring to share the spectrum with the concerned primary users. If P_i is the tariff/pricing policy and the QoS guaranteed by primary user i then each of the secondary subscribers strives to subscribe at the given tariff so as to attain a QoS sufficient to satisfy individual needs. The secondary users utilize adaptive modulation for transmissions in the allocated spectrum in a time-slotted manner. In this kind of modulation, transmission rate is a function of channel quality, while bit error rate must be maintained at specified target levels.

Accordingly, the spectral efficiency of transmission for secondary user i can be expressed as:

$$k_i = log_2(1 + Ky_i)$$

where

$$K = \frac{1.5}{\ln\left(\frac{0.2}{BER_i^{tar}}\right)}$$

The secondary user i transmits with spectral efficiency k_i to the extent that the demand of the secondary users is a function of transmission rate in the allocated frequency spectrum as well as the price charged by the primary users.

Table 1   Notation summary

Symbols	Description
λ_i	Arrival rate
Q_i	Spectrum size (Secondary user)
W_i	Spectrum size (Primary user)
$p^{(i)}$	Price
P_j	Price
$k_i^{(p)}$	Spectral efficiency (Primary users)
$k_i^{(s)}$	Spectral efficiency (Secondary users)
c_i^D	Cost function (delay)
c_i^T	Cost function (Throughput)
d_i	constant (elasticity)
D_i	Delay
ψ	Utility
(Q)	Set of available spectrum size
Δ	Substitutability
$\phi_i(T)$	Profit (Throughput)
$\phi_i(D)$	Profit (Delay)
y_i	Channel quality (player i)
y_j	Channel quality (player j)
T	Throughput
n	number of users
β	constant

QoS Measure and Cost

The QoS performance of a primary user is degraded in the event of some portion of spectrum being shared with the secondary user. Thus cost function must be considerate of the QoS performance of the primary user. On this basis we consider a two pronged QoS measure. The first one is average delay as a QoS measure obtained for the transmissions at the primary user based on an M/D/1 queueing model [18] Throughput Measure Regarding the delay QoS measure, is defined as:

$$D_i(Q_i) = \frac{1}{2} \frac{\lambda_i}{\left(k_i^{(p)}(W_i - Q_i)^2 - \lambda_i k_i^{(p)}(W_i - Q_i) \right)}$$

with the symbols meaning as given in the table, it is worth to note that $k_i^{(p)}(W_i - Q_i)$, denotes the service rate. The cost function is defined as:

$$C_i^D = dD_i(Q_i)$$

The other QoS measure is the throughput given by:

$$T(Q_i) = \sum_{i=1}^{N} \frac{\beta Q_i}{\sqrt{nlogn}}$$

The cost due to this measure is expressed as:

$$C_i^T = dT_i(Q_i)$$

Utility Function

The utility gained by the secondary users makes it possible to ascertain the level of spectrum demand. A quadratic utility function defined as in [17]:

$$\psi(Q) = \sum_{i=1}^{M} Q_i k_i^s - \frac{1}{2}\left(\sum_{i=1}^{M} Q_i^2 + 2\Delta \sum_{i=1}^{M} Q_i Q_j \right) + J$$

where $Q = Q_1, ..., Q_i..., Q_M$ and J is given by:

$$J = -\sum_{i=1}^{M} P_i Q_i$$

The spectrum substitutability is included in the utility function by way of parameter ∇. This parameter permits the secondary users to switch between frequencies depending on the offered price. The demand function of the secondary user is obtainable from differentiating the utility function w.r.t Q_i as follows:

$$\frac{d\psi(Q)}{dQ_i} = 0$$

The demand function is the size of shared spectrum that maximizes the utility of the secondary user given the prices offered by the primary service

$$Q_i = \frac{k_i^{(s)} - p_i - \Delta(k_j^{(s)} - p_j)}{1 - \Delta^2}$$

B. Bertrand Game Model

The Bertrand oligopoly is formulated as in Table 2. The profit due to a delay QoS performance is:

$$\phi(P)_i^{(D)} = Q_i P_i - C_i^{(D)}$$

While the throughput based profit is

$$\phi(P)_i^{(T)} = Q_i P_i - C_i^{(T)}$$

The solution to this game is the Nash Equilibrium (NE), obtainable by way of the best response. For a best response of a Primary user i given the prices of other primary users P_i, where $j \neq i$ is defined as

$$BRi(P_{-i}) = \arg\max \phi_i\ (P_{-i} \cup P_i)$$

The set $P^* = \{P_1^*,..., P_N^*\}$ represents the nash equilibrium of this Bertrand game, if and only if

$$P_i^* = BRi(P_{-i}^*), \forall i$$

The *NE* value in the context of delay QoS measure is obtainable by differentiating

$$\frac{d\phi(Q)}{dP_i} = 0$$

for all i where

$$\phi(P) = Pi\frac{k_i^s - P_i - \nabla(k_j^{(s)} - P_j)}{1 - \nabla^2}$$

$$- \frac{d\lambda_i}{2(W_i - Q_i)^2 - 2\lambda_i(W_i - Q_i)}$$

Table 2   Bertrand game formulation

Entity	Description
Players	Primary users
Strategies	Price per unit of spectrum (P_i)
Payoffs	The payoff for each player is the profit of primary user

The derivative of this profit function is equated to zero as follows

$$0 = \frac{k_i^{(s)} - 2P_i - \Delta(k_i^{(s)} - P_j)}{1 - \Delta^2} + \frac{d\frac{\lambda_i}{1-\Delta^2(4Q_i-\lambda_i)}}{(2Q_i^2 - 2Q_i\lambda_i)^2}$$

$$Q_i = W_i - \frac{k_i^{(s)} - Pi - \Delta(k_j^{(s)} - P_j)}{1 - \Delta^2}$$

We further extend the our efforts to encompass the QoE in the context of a low cost Smart mesh network using the formular in [8] coupled with the delay and throughput equations.

4 Performance Evaluation

A. Parameter Setting

The parameters are set as in Table 3

B. Numerical Analysis

In this section, we present numerical results to validate the efficacy of our low cost Smart Mesh network design using the two analytic models.

Figure 2 depicts the demand function of the secondary user, the revenue, cost and profit of the primary user under variable pricing options for the delay and throughput QoS performance metrics respectively. From a delay QoS performance metric perspective, when the first primary user strategizes by increasing the spectrum price, the secondary user correspondingly demands

Table 3 System parameters

Parameter	Value
Primaryuser Spectrum	5MHz
BER	10^{-4}
Traffic Arrival Rate	1Mbps
d	1
Channel Quality Span	10–20dB
λ_i	4
y_1	15
y_2	18
Δ	0.4
P_2	1
Primaryusers	2

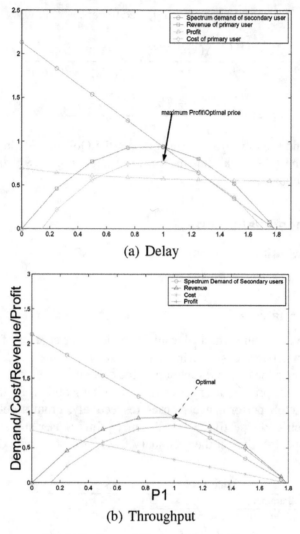

(a) Delay

(b) Throughput

Figure 2   Demand, Revenue, Cost and Profit

less spectrum owing to the decrease in the utility of the allocated spectrum. Moreover, the cost for the primary user decreases given a small demand from the secondary user. Needless to say, the size of the residual spectrum remains bigger giving rise to a small delay. However the revenue and profit of the primary user, traverses a parabolic path as it initially increases and then after the optimal point begins to decrease. Clearly for a small price, the first primary user can sell a bigger spectrum size to the secondary user, this translates to an increase in revenue and profit. Comparatively from a throughput QoS performance metric perspective, when the spectrum price increases, little spectrum is sold. Similarly when the primary user increases the price, the secondary user correspondingly demands less spectrum and vice-versa. However, the cost function shows a cost that is initially higher than that in the delay metric and then decreases sharply with an increase in price as depicted by the negative line gradient in the throughput version of the graph. The revenue and profit functions also follow a parabolic path. Notably for the two QoS constraints, there exist points of maximized profit at which the price is considered optimal. The gap between the two parabolic curves, i.e., profit curve and revenue curve is in a way reflective of the differences in the cost functions.

In Figure 3, we consider two primaries and their best responses under the delay and throughput QoS constraints. This in a way depicts attempts to catalyze spectrum price decrease and a subsequent increased access to internet services. The price catalyzation is brought about by a change in strategy by both Primary 1 and Primary 2 as they both seek to attain the best price that will be attractive to the secondary user. The price strategy is itself a function of channel quality, thus when channel quality increases, the spectrum demand increases as it gives the secondary user a higher rate due to adaptive modulation. Consequently in accordance with the law of demand and supply in economics, the primary user sets a higher price. The intersection of the best response lines from both primary 1 and primary 2 depicts the location of the optimal point which is also the Nash equilibrium point. The Nash equilibrium points for the delay metric are located at a lower position value points as compared to those of the throughput performance metric. This intuitively means it may it advisable to employ this performance metrics in attempts to catalyze a decrease in service prices and subsequently enable entrepreneurs to achieve increased access in the rural and remote parts. Next we investigate and anlayze Nash equilibrium under variable channel quality depicted by Figure 4 for both performance metrics. A higher channel quality is deliverable via the delay QoS metric as compared to its throughput counterpart. This translates to a higher Nash equilibrium point for the delay QoS metric. This is a result of a higher demand emanating from the secondary users. For both graphs and metrics, the channel quality offered by one primary impacts the strategies adopted by the other primary. Consequently when the demand offered by one player is varied, the other player must responsively adopt the price to attain higher price. Utimately, the throughput delivers the same channel quality at

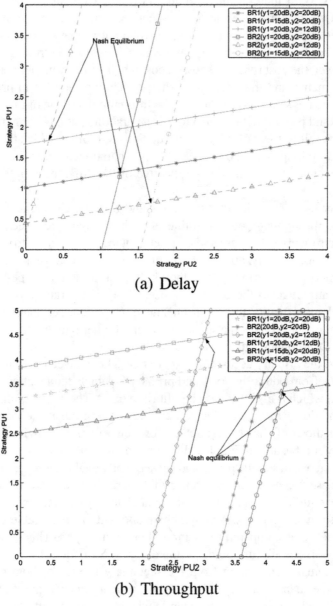

(a) Delay

(b) Throughput

Figure 3   Best response

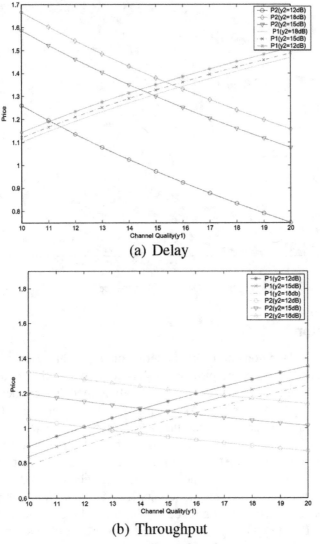

(a) Delay

(b) Throughput

Figure 4 Channel Quality

a decreased price, a fact which gives the throughput based model an edge over the delay based model. The choice of a throughput based model is also confirmed by the QoE graph in Figure 5. The top graph depicts a predicted user perception of the throughput model while the bottom shows the delay model perception.

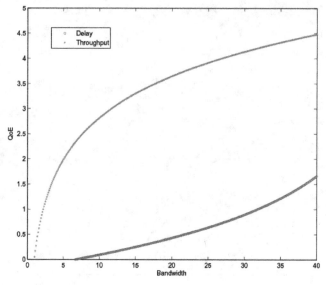

Figure 5   Quality of Experience

5 Conclusion

This paper studied the non cooperative interaction of primary users (licensed users) and secondary users (employing mesh routers) within the context of a smart mesh network. Two non -cooperative analytic models were developed for a TV white space spectrum market applicable in rural and remote areas by entrepreneurs when provisioning internet access via smart wireless mesh network. The models are based on the delay and throughput QoS performance metrics. Objectively the models strive to catalyze a decrease in costs (prices) and increase broadband internet access. The throughput based model is according to our performance evaluation superior at delivering high quality at a decreased cost price as compared to the delay based model. This is further substantiated by QoE prediction. Further work could involve the use of different utility functions and applying these models to a cognitive routing scenario in which suitable routes are selected based on an adequate strategy.

References

[1] D. Niyato and E. Hossain. *Competitive Pricing in Heterogeneous Wireless Access Networks: Issues and Approaches*, in IEEE Network, IEEE, vol.22, no.6, pp. 4, 11, November-December 2008.

[2] K. R. Chowdhury and I. F. Akyildiz. *Cognitive Wireless Mesh Networks with Dynamic Spectrum Access*, in Selected Areas in Communications, IEEE Journal on, vol. 26, no.1, pp. 168, 181, January 2008.

[3] I. F. Akyildiz, W. Y Lee, M. C. Vuran and S. Mohanty. *NeXt generation/dynamic spectrum access/cognitive radio wireless networks: A survey*, in Computer Networks, Volume 50, Issue 13, pp. 2127–2159, 15 September 2006.

[4] P. Ballon. *Changing Business models for Europe's mobile telecommunications industry: The impact of alternative wirelss technologies* , in Elsevier Telematics and Informatics Journal, Volume 24, number 3, pp. 192–2005, 2007

[5] *Dynamic Spectrum Access (DSA) for Economical Rural Broadband Internet*, retrieved (http://www.sharedspectrum.com/wp-content/uploads /090814-Rural-Broadband-White-Paper-v1-3.pdf)

[6] C. McGuire, M. R Brew, F. Darbari, S. Weiss, R. W Stewart. *Enabling rural broadband via TV White Space*, in 5th IEEE International Symposium on Communications Control and Signal Processing (ISCCSP), 2–4 May 2012.

[7] S. Song. *Spectrum for development: The Impact of Access and The Role of Wireless Technologies, October 2011*, Retrieved http://www.apc.org/ en/system/files/OpenSpectrumIssuePaper-EN.pdf

[8] J. Shaikh, M. Fielder and D. Collange. *Quality of Experience from user and network perspectives*, in Springer Annals of Telecommunication journal, Vol 65, number 1–2, pages 47–57, 2010

[9] S. M Nleya, A. Bagula, M. Zennaro, E. Pietrosemoli. *A TV White Space Broadband Market Model for Rural Entrepreneurs*, in 5th Global Information Infrastructure and Networking Symposium, 28–31 October 2013

[10] Y. Xing, R. Chandramouli, and C. M. Cordeiro. *Price dynamics in competitive agile spectrum access markets*, in IEEE Journal on Selected Areas in Communications, vol. 25, no. 3, pp. 613–621, April 2007.

[11] S. Gandhi, C. Buragohain, L. Cao, H. Zheng, and S. Suri. *A general framework for wireless spectrum auctions*, in Proc. IEEE International Symposium on New Frontiers in Dynamic Spectrum Access Networks (DySPAN07), April 2007.

[12] Y. Xing, R. Chandramouli, and C. M. Cordeiro. *Price dynamics in competitive agile spectrum access markets*, in IEEE Journal on

Selected Areas in Communications, vol. 25, no. 3, pp. 613–621, April 2007

[13] D. Niyato and E. Hossain. *A game-theoretic approach to competitive spectrum sharing in cognitive radio networks*, in Proc. IEEE Wireless Communications and Networking Conference (WCNC07), March 2007

[14] M. Thoppian, S. Venkatesan, R. Prakash, and R. Chandrasekaran. *MAC-layer scheduling in cognitive radio based multi-hop wireless networks*, in Proc.WoWMoM06, June 2006.

[15] D. Niyato, E. Hossain, Le Long. *Competitive Spectrum Sharing and Pricing in Cognitive Wireless Mesh Networks*, in Wireless Communications and Networking Conference, 2008. WCNC 2008. IEEE, vol., no., pp. 1431, 1435, March 31 2008–April 3 2008

[16] F. Fang, Q. Lili and A. B. Whinston. *On Profitability and Efficiency of Wireless Mesh Networks*, in 15th Annual Workshop on Information Technologies and Systems (WITS), March 2006

[17] N. Singh and X. Vives. *Price and Quantity Competition in a differential duopoly*, in RAND Journal of Economics, vol.15, no. 4, pp. 546–554, 1984

[18] D. Niyato and E. Hossain. *Microeconomic Models for Dynamic Spectrum Management in Cognitive Radio Networks*, invited chapter in Cognitive Wireless Communication Networks, (Eds. V. K. Bhargava and E. Hossain), Springer-Verlag, November 2007.

Biographies

Sindiso M Nleya received the BSc degree in Applied Physics and the MSc degree in Computer Science from the National university of Science and Technology (NUST), Bulawayo, Zimbabwe, in 2003 and 2007, respectively. In 2008 he joined the Computer Science department in the same university as a member of academic staff. He is currently pursuing a PhD in computer

Science at the University of Cape Town, South Africa and is a member of the Intelligent systems and Advanced Telecommunications laboratory. His research interests focus on Dynamic Spectrum Access, Algorithmic game theory and Optimization Techniques.

Antoine Bagula obtained his doctoral degree from the KTH-Royal Institute of Technology in Sweden. He held lecturing positions at the University of Stellenbosch and the University of Cape Town before joining the Computer Science department at the University of the Western Cape in January 2014. Professor Bagula has been on the technical programme committees of more than 50 international conferences and on the editorial board of international journals. He has also co-chaired international conferences in the field of telecommunications and ICT. Professor Bagula has authored/co-authored more than 100 papers in peer-reviewed conferences and journals and book chapters. Bagula's research interest is Computer Networking with a specific focus on the Internet-of-Things, Cloud Computing, Network security and Network protocols for wireless, wired mesh networks and hybrid networks.

Marco Zennaro is a researcher at the Abdus Salam International Centre for Theoretical Physics in Trieste, Italy, where he coordinates the Telecommunications/ICT4D Laboratory. He received his PhD from the

KTH-Royal Institute of Technology, Stockholm, and his MSc degree in Electronic Engineering from the University of Trieste. His research interest is in ICT4D, the use of ICT for Development, and in particular he investigates the use of wireless sensor networks in developing countries. Dr. Zennaro is one of the authors of "Wireless Networking in the Developing World," which has been translated in six languages.

Ermanno Pietrosemoli is a researcher at the Telecommunications/ICT for Development Laboratory of the Abdus Salam International Centre for Theoretical Physics in Trieste, Italy, and president of Fundación Escuela Latinoamericana de Redes "EsLaRed", a non-profit organization that promotes ICT in Latin America through training and development projects. EsLaRed was awarded the 2008 Jonathan B. Postel Service Award by the Internet Society. Ermanno has been deploying wireless data communication networks focusing on low cost technology, and has participated in the planning and building of wireless data networks in Argentina, Colombia, Ecuador, Italy, Lesotho, Malawi, Mexico, Morocco, Nicaragua, Peru, Senegal, Spain, Trinidad, U.S.A., Venezuela and Zambia. He has presented in many conferences and published several papers related to wireless data communications. He is one of the authors of the book "Wireless Networking in the Developing World".

Ermanno holds a Master's degree from Stanford University and was a professor of Telecommunications at Universidad de los Andes in Venezuela from 1970 to 2000.

Performance Analysis of RoFSO Links with Diversity Reception for Transmission of OFDM Signals Under Correlated Log-normal Fading Channels

Fan Bai, Yuwei Su and Takuro Sato

Graduate School of Globe Information and Telecommunication Studies, Waseda University, Tokyo 169-0072, Japan
E-mail: baifan@ruri.waseda.jp, bruce.suyuwei@yahoo.co.jp, t-sato@waseda.jp

Received: October 14, 2014;
Accepted: November 10, 2014

Abstract

Free space optical (FSO) communication has been receiving growing attention with recent commercialization successes as a cost-effective and high bandwidth optical access technique. Meanwhile, FSO communication has been regard as an attractive solution to bridging the gap between the wireless communications and optical fiber communications. However, a significant performance degradation in FSO communication system due to the atmospheric turbulence impairs the transmission performance improvement. FSO system employing the spatial diversity technique can be used to mitigate the effect of turbulence and improve the transmission performance. In this paper, a novel analytical approach is presented to evaluate the transmission performance of OFDM-FSO system with diversity reception considering effect of channel correlation. A detailed mathematical model for OFDM-FSO system over turbulent channel modeled by correlated Log-normal distribution is provided. Then, We derive the signal-to-noise ratio (SNR), bit error ratio (BER) and outage probability expressions taking into account the diversity combining schemes (i.e. MRC, EGC), effects of atmospheric turbulence, channel correlation and aperture size of receiver lens. The results of this

Journal of ICT, Vol. 2, 129–150.
doi: 10.13052/jicts2245-800X.224

study show that the most significant parameters that degrade the system performance. Furthermore, the obtained numerical results can be useful for designing, evaluating and enhancing the FSO system's ability to transmit wireless signal under actual conditions.

Keywords: Free Space Optical (FSO), Orthogonal Frequency Division Multiplexing (OFDM), spatial diversity, channel correlation, atmospheric turbulence, Maximum Ratio Combining (MRC), Equal Gain Combining (EGC).

1 Introduction

Recently, with the growing demand on the high-speed and high-quality applications for the communication devices such as high-definition television (HDTV), video programs and so on, we observe an explosion in the network traffic. Free space optical (FSO) system known as optical wireless communication (OWC) application was proposed for its potential high data rate capacity, low cost, and wide bandwidth on unregulated spectra. At the same time, transmission of radio frequency (RF) signals by means of optical fiber links, which are commonly referred to as radio over fiber (RoF), has been utilized for many years as a cost effective and high-capacity solution to connect the wireless signal and optical fiber communication [1]. FSO link can conveniently be used to transmit RF signals, which are similar to RoF but excluding the fiber medium as radio-on free space optics (RoFSO) [3]. Thus, the commercial FSO systems can be seem as a good candidate for supporting the heterogeneous services in the next generation access networks. In the mean time, ITU-T also published the first ITU-T recommendation in the area of G.640 for FSO application in the practical issue [2].

For the case of FSO link model, one of the main problems that affect the transmission performance is irradiance fluctuations induced by the atmospheric turbulence. This irradiance known as the optical scintillation, can cause the power losses [5]. It is the major performance impairment in FSO system. Therefore, it is importance to apply efficient techniques to mitigate the channel fading caused by atmospheric turbulence.

Spatial diversity is an efficient solution to mitigate channel fading by using the multiple apertures at the receiver (SIMO) or transmitter side (MISO) or combination (MIMO) [9]. Further, for the 5G communication technology, massive MIMO is regard as one of the important solution. However, diversity techniques are most efficient under the conditions of uncorrelated fading on the sub-channels [6, 10]. In the previous studies, any two adjacent antennas must be sufficiently far apart as independent case due to channel correlation will impair the transmission performance improvement [7]. Therefore, one of the key points is determining the relation between sub-channel correlation among each adjacent antenna over the long transmission distance and transmission performance of FSO system.

Orthogonal frequency division multiplexing (OFDM) is a widely applied technique for wireless communications that divides the spectrum into a number of equally spaced sub-channels and carries a portion of a users information on each channel [5]. OFDM/OFDMA, which is the air interface for the Long Term Evolution (LTE) in wireless communication systems, is regard as one of important broadband application areas using RoF technology. Meanwhile, OFDM has been adopted in several high-speed digital communication standards such as digital terrestrial TV broadcasting and the IEEE 802.11 local area network (LAN) and IEEE 802.16 standards [8].

Furthermore, in order to get an insight into the atmospheric channel statistical models effects to the system performance, we consider the correlated Log-normal distribution due to its excellent match to multiple sub-channel links over weak turbulence fading channel. Since the FSO link can be modeled as diversity reception, an efficient diversity combining scheme is required to improve the system performance. As we know, maximum ratio combining (MRC) can provide an optimum combining performance, but implementation complexities are inherent, and system is extremely sensitive to channel estimation error. Equal gain combining (EGC) is inefficient for system with branches having acutely low SNR conditions [8, 9]

In this paper, we propose a OFDM-FSO system with dual diversity reception over correlated Log-normal distribution fading channel. The goal of this paper is to analyze and explore the potential of OFDM-FSO system with diversity reception for use in high-performance high-speed transmission. Then, we investigate channel correlation effects on transmission performance, in terms of averaging bit error ratio (BER), signal-to-noise ratio (SNR), and outage probability (OP). We take into consideration the effect of diversity combining scheme, aperture size of receiver lens and number of diversity reception. In this analysis, the use of spatial reception configuration with OFDM modulation can be seem as a countermeasure for the mitigation of scintillation and enhancing the performance of FSO systems in operation environments. Besides, our theoretical study provides guidelines to optimally configure the actual system design by applying an optimum transmission performance.

The rest of this paper is organized as follows. In Section 2, we derive the expressions for the channel correlation coefficient for a dual diversity reception based on the plane wave model. Then, we model the probability density function of bivariate Log-normal random variables based on diversity reception in Section 3. In the Section 4, we introduce the dual diversity OFDM-FSO system structure and present the mathematical modeling for the transmission of OFDM signal over fading channel modeled by correlated Log-normal distribution. In the Section 5, we provide the numerical results to analyze and discuss the effect of the spatial diversity on the OFDM-FSO link over turbulent channel. Finally, we give the conclusion in Section 6.

2 Channel Correlation in Atmospheric FSO Links with Diversity Reception

In this section, we review the theories used to derive the expression for the channel correlation for diversity reception over turbulent channel. Then, we present the mathematical modeling for the joint probability density function (PDF) which represents two Log-normal random variables (RVs) of the light intensity by turbulence channel.

The most serious effects that severely affect the propagation of a laser beam propagating through the atmosphere are caused by variations of the refractive index in turbulent channel [4]. An important parameter that describes the strength of turbulence is the scintillation index: $\sigma_I^2 = \langle I^2 \rangle / \langle I \rangle^2 - 1$, with I being the intensity of optical wave. In fact, atmospheric turbulence as an important influence factor which it can cause the intensity fluctuation in the received signal level and leads to sharp increase in the bit error in a FSO communication link. Intensity fluctuation at a received power variance that depends on the size of the receive aperture, most of previous literatures proposed reduce the power variance by increasing the diameter of aperture as aperture averaging (AA) method [5, 6]. Aperture averaging can be seen as a simple form of spatial diversity when the receiver lens aperture is larger than the fading correlation length [8].

In this paper, we used spatial domain technique model a simple dual diversity reception FSO system and collect the OFDM signal at different two positions. Consider a dual diversity reception FSO link transmitted by plane wave model as show in Figure 1. We define the channel correlation coefficient ρ_{12} between any component receivers, denoted 1*th* receiver and 2*th* receiver:

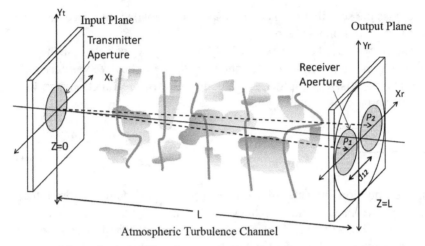

Figure 1 A dual diversity reception based on plane wave model

$$\rho_{12}\left(d_{12}, D\right) = \frac{B_{I,12}\left(I_1, I_2, d_{12}, D\right)}{\sqrt{\sigma_1^2 \sigma_2^2}} \tag{1}$$

with d_{12} being the separation distance of two points in the adjacent receiver aperture plane, D is aperture diameter. In this paper, d_{12} is a center-to-center separation distance. $B_{I,12}$ and σ_j^2 $(j = 1, 2)$ are spatial covariance function of the irradiance and scintillation index of the jth diversity receiver. In order to characterize the irradiance and the scintillation index of receiver induced channel fading, we use the light irradiance spatial co variance function [5]:

$$B_{I,12}\left(I_1, I_2, d_{12}, D\right) = \frac{\langle I_1 I_2 \rangle}{\langle I_1 \rangle \langle I_2 \rangle} - 1 \tag{2}$$

$$B_{I,i,j}\left(I_i, I_j, 0, D\right) = \sigma_{i=j}^2\left(D\right) = \frac{\langle I_j^2 \rangle}{\langle I_j \rangle^2} - 1 \tag{3}$$

where $< . >$ for ensemble average over different turbulence disorder and I_j is the light intensity collected by the jth receiver. Previous equations indicate that the irradiance spatial covariance function is a more general statistic model that includes the scintillation index as a special case when the space distance between adjacent point as zero $(d_{12} = 0)$.

Furthermore, based on the extended Rytov theory, we can also use the log-amplitude X and log-amplitude variance σ_X^2 of the optical wave represents two points on-axis log-amplitude fluctuation covariance function, the log-amplitude variance is defined by [5]

$$
\begin{aligned}
B_{X,12}(r_1, r_2, L) &= \left\langle (X_1 X_2)^2 \right\rangle - \left\langle (X_1 X_2) \right\rangle^2 \\
&= \tfrac{1}{2} Re[E_2(r_1, r_2, L) + E_3(r_1, r_2, L)]
\end{aligned}
\tag{4}
$$

when the equation indicates the on the optical axis point log-amplitude fluctuation covariance means that the valued of the transverse vector at output plane with $r_j = 0$. Where $Re(.)$ denotes the real part of X, and E_2, E_3 are two second-order statistical moments given by [5, 13]

$$
\begin{aligned}
E_2(0, 0, L) &= 4\pi^2 k^2 L \int_0^1 \int_0^\infty k \Phi_{n.eff}(k) \\
&\quad exp\left(-\frac{k^2 D^2}{16}\right) J_0(K\, d_{12}) dk d\xi
\end{aligned}
\tag{5}
$$

$$
\begin{aligned}
E_3(0, 0, L) &= -4\pi^2 k^2 L \int_0^1 \int_0^\infty k \Phi_{n.eff}(k) \\
&\quad exp\left(-\frac{k^2 D^2}{16} - i\frac{L\xi}{k}\right) J_0(K\, d_{12}) dk d\xi
\end{aligned}
\tag{6}
$$

where $\xi = 1 - z / L$ and $J_0(.)$ is Bessel function of the first kind and zero order, k is optical wave number. The L represent transmission distance. The effective atmospheric spectrum $\Phi_{n.eff}(k)$ is used so that the covariance function is valid under all turbulence conditions. When the effective atmospheric spectrum $\Phi_{n.eff}(k)$ under neglect inner-scale and outer-scale situation is defined as [4]

$$
\Phi_{n.eff}(k) = 0.033 C_n^2 k^{-11/3} \left[exp\left(-\frac{k^2}{k_{X.0}^2}\right) + \frac{k^{11/3}}{(k^2 + k_{Y.0}^2)^{11/6}} \right]
\tag{7}
$$

with

$$
k_{X.0}^2 = \frac{k}{L} \frac{2.61}{1 + 1.1\sigma_R^2}
\tag{8}
$$

$$k_{Y,0}^2 = \frac{3k}{L}(1 + 0.69\sigma_R^{12/5}) \tag{9}$$

The $k_{X,0}$ and $k_{Y,0}$ represent the low-pass and high-pass spatial frequency cutoffs, σ_R^2 for turbulence strength. When the log-amplitude variance is sufficiently small, and making $r_1 = r_2 = 0$ for the on-axis values, the scintillation index is related by

$$B_{I,jj}(I_j, I_j, 0, D) = \sigma_j^2(D) = exp\,[4B_{X,12}(r_1, r_2, L)] - 1 \tag{10}$$

Then, when the observation points on the optical axis, the equation (2) can be changed to

$$B_{I,12}(I_1, I_2, d_{12}, D) = exp\{2\,\mathrm{Re}[E_2(0, 0, L) + E_3(0, 0, L)]\} - 1$$
$$= exp\left\{8\pi^2 k^2 L \int_0^1 \int_0^\infty \Phi_{n.eff}(k)exp(-\frac{k^2 D^2}{16})\right.$$
$$\left. \times\, J_0(kd_{12}) \left[1 - cos\left(\frac{Lk^2\xi}{k}\right)\right] dkd\xi \right\} - 1 \tag{11}$$

3 Correlated Log-normal Distribution Model

Theory reliably of FSO system operating in presence of atmospheric turbulence can be based on a mathematical model of the probability density function of randomly light intensity irradiance that describes the system's behavior across the different turbulence strength regimes [6]. In this paper, we model the correlated Log-normal distribution represents combine faded of intensity irradiance, then, we investigate the effect of channel correlation to system performance under weak turbulence regime. Based on previous literatures [5, 13], which single receiver Log-normal probability density function (PDF) of light intensity fading is given by

$$f_I(I) = \frac{1}{2I}\frac{1}{(2\pi\sigma_X^2)^{1/2}}exp\left\{-\frac{(\ln(I) - \ln(I_0))^2}{8\sigma_X^2}\right\} \tag{12}$$

where I_0 denotes the light intensity in the absence of turbulence, $I = I_0 exp\,(2X - 2\langle X\rangle)$. Hence, a diversity reception FSO system where optical signals collected by multiple apertures undergo diversity combining, we model the spatial covariance matrix K as

$$K = \begin{bmatrix} k_{11} & k_{12} & \cdots & k_{1n} \\ k_{21} & k_{22} & \cdots & k_{2n} \\ \vdots & \vdots & \ddots & \vdots \\ k_{n1} & k_{n2} & \cdots & k_{nn} \end{bmatrix} \tag{13}$$

where $k_{ij} = B_{I,ij}$ is spatial covariance function, $B_{I,ij} = \sigma_j^2(D)$ when the $i = j$ as equation (2) and (3). This spatial covariance matrix can be represented multiple transceivers case to determine the effect of correlation among the sub-channels. Then, we assume correlated Log-normal distribution function of the light intensity of dual sub-channels is given by

$$f_{I_1,I_2}(I_1, I_2) = \frac{1}{2^2 I_1 I_2} \frac{1}{2\pi \sigma_{X1} \sigma_{X2} \sqrt{1 - \rho_{12}^2}}$$
$$\times exp\left\{ -\frac{1}{8(1 - \rho_{12}^2)} \left[\frac{(lnI_1 - lnI_{1,0})^2}{\sigma_{X1}^2} + \frac{(lnI_2 - lnI_{2,0})^2}{\sigma_{X2}^2} \right. \right.$$
$$\left. \left. -2\rho_{12} \frac{(\ln I_1 - lnI_{1,0})(lnI_2 - lnI_{2,0})}{\sigma_{X1}\sigma_{X2}} \right] \right\} \tag{14}$$

where $I_{1,0}$ and $I_{2,0}$ are the received light intensity at the $1th$ receiver and $2th$ receiver in the absence of turbulence situation.

4 Performance Analysis of a Atmospheric OFDM-FSO Link with Diversity Reception

In this paper, the main focus is to establish an analytical model able to characterize the OFDM signal transmission through an intensity modulation/direct-detection (IM/DD) FSO link. We derive the expressions of the average signal-to-noise ratio (SNR), average bit error ratio (BER) and outage probability, which analysis considers on the effect of scintillation, both of MRC and EGC combining schemes and aperture size of receiver lens in this section. Then, a dual diversity reception OFDM-FSO system is considered where the OFDM signal is transmitted via single aperture and collected by two apertures ($j=2$) over turbulent channel modeled by correlated Log-normal distribution with additive white Gaussian noise (AWGN). Figure 2 shows OFDM-FSO diversity receivers system architecture. The OFDM signal for N_s sub-carriers, after up-conversion can be written as

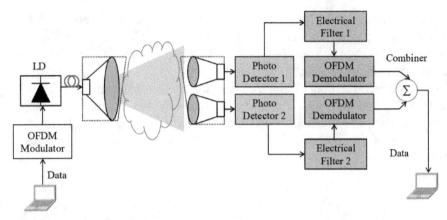

Figure 2 OFDM-FSO system with dual diversity reception over atmospheric turbulence channel

$$S_{OFDM}(t) = \sum_{n-0}^{N_s-1} X_n exp\{j(\omega_n + 2\pi f_c)t\}, 0 \leq t < T_s \quad (15)$$

Where ω_n are the set of orthogonal sub-carriers frequency, T_S is the OFDM symbol duration, $X_n = a_n + jb_n$ is complex data symbol in the nth subcarrier, with an and bn the in-phase and quadrature modulation symbols. The first raw data is mapped according to different types of modulation techniques. Each symbol X_n is amplitude modulated on orthogonal sub-carriers. The signal $S_{OFDM}(t)$ is then used to modulate the optical intensity of laser-diode (LD) to be transmitted through the fiber optics. In this paper, we neglect the inter-modulation distortion (IMD). Then, the optical power output from the LD can be expressed as

$$P(t) = P_t \left(1 + \sum_{n=0}^{N-1} m_n S_{OFDM}(t)\right) \quad (16)$$

where P_t is average transmitted optical power, m_n is optical modulation index on per sub-carrier. At the receiver side, the received optical power in the presence of turbulence at the photo detector input can be express as

$$P_{r.FSO}(t) = P(t)L_{loss} X + n_{FSO}(t) \quad (17)$$

where L_{loss} is the sum of losses due to atmospheric turbulence, geometrical loss and pointing error. The $n_{FSO}(t)$ characterizes the additive white Gaussian noise (AWGN), X quantifies the variation of the signal fading due to atmospheric turbulence effects.

Then, the received optical signal convert to the electrical signal by photo-detector (PD), and the FSO noise can be filtered in the PD. Then, signal pass through the electrical filter and OFDM demodulator. Assuming the total output photo current of the combiner in the presence of turbulence can be expressed as follow

$$i(t, I_1, I_2, X) = I_{ph}(I_1, I_2) \left(1 + \sum_{n=0}^{N-1} m_n S_{OFDM, 1, 2}(t) \right) + n_{opt, 1, 2}(t)$$
(18)

where $I_{ph} = R_D P_t L_{loss} X$ is dc of the received photo current depend on received light intensity, and including I_1 and I_2 received by each receiver, R_D is detector responsivity, m_n is optical modulation index on per subcarrier, and $n_{opt}(t)$ is the sum of thermal noise, shot noise and is modeled as Gaussian white random process with zero mean and variance $\sigma_N^2 = N_0/2$. The total noise power is given by

$$N_0(I_1, I_2) = \frac{4K_B T_{abs} F_e}{R_L} + 2qI_{ph}(I_1, I_2) + (RIN)I_{ph}^2(I_1, I_2)$$
(19)

where K_B is the Boltzmanns constant, T_{abs} is the absolute temperature, F_e is the noise figure of the receiver electronics, RIN is relative intensity noise, R_L is the PD load resistor, and q is the electron charge.

In this analysis, we compare the performance of dual diversity reception with two diversity combining schemes, maximal ratio combining (MRC) and equal gain combining (EGC). In the presence of turbulence, the instantaneous electrical signal-noise-to ratio (SNR) at the output of the diversity combiner using MRC and EGC are given by [13]

$$\gamma_{total.MRC}(I_1, I_2, X_1^2, X_2^2) = \sum_{j=1}^{J} \gamma_j(X_j^2), (j = 2)$$
(20)

$$\gamma_{total.EGC}(I_1, I_2, X_1^2, X_2^2) = \left(\sqrt{\gamma_1(X_1^2)} + \sqrt{\gamma_2(X_2^2)} \right)^2 / 2$$
(21)

where γ_j is instantaneous SNR of *j*th receiver in the turbulent channel. Thus, the received photo current is proportional to *X* and has same statistic in the presence of the turbulence. Moreover, consider a dual diversity reception OFDM-FSO system using M-QAM modulation where the sub-channel collected light intensity (I_1, I_2) follow a correlated Log-normal distribution. Taking into consideration the correlation between the receivers with log-amplitude fluctuation and represented by *X*, thus, the equation (14) can be modified as follow

$$f_{X1,X2}(X1, X2) = \frac{1}{2\pi\sigma_{X1}\sigma_{X2}\sqrt{1-\rho_{12}^2}}$$
$$\times exp\left\{-\frac{1}{2(1-\rho_{12}^2)}\left[\frac{(X_1-\langle X_1\rangle)^2}{\sigma_{X1}^2} + \frac{(X_2-\langle X_2\rangle)^2}{\sigma_{X2}^2}\right.\right.$$
$$\left.\left.-2\rho_{12}\frac{(X_1-\langle X_1\rangle)(X_2-\langle X_2\rangle)}{\sigma_{X1}\sigma_{X2}}\right]\right\} \quad (22)$$

Finally, the total average BER for the received QAM-OFDM signal, where $M = 2^n$ and *n* is an even number, is given by [13]

$$\langle BER_{total}\rangle = \int_0^\infty \int_0^\infty f_{X_1,X_2}(X_1, X_2)BER_{total}(X_1^2, X_2^2)dX_1dX_2 \quad (23)$$

where instantaneous electrical BER_{total} is

$$BER_{total}(X_1^2, X_2^2) = \frac{2(1-\sqrt{M}^{-1})}{\log_2^M}er fc\left(\sqrt{\frac{3\gamma_{total}(X_1^2, X_2^2)}{2(M-1)}}\right) \quad (24)$$

The case of MRC reception combining scheme with dual diversity reception, the electrical SNR in the presence of turbulence can be written as [13]

$$\gamma_{total.MRC}(X_1^2, X_2^2) = \frac{0.5m_n^2(I_{ph.1}^2(I_1, X_1^2) + I_{ph.2}^2(I_2, X_2^2))}{N_0B_e} \quad (25)$$

and EGC reception combining scheme can be modeled by [13]

$$\gamma_{total.EGC}(X_1^2, X_2^2) = \left[\frac{\sqrt{0.5}m_n(I_{ph.1}^2(I_1, X_1^2) + I_{ph.2}^2(I_2, X_2^2))}{\sqrt{2N_0B_e}}\right]^2 \quad (26)$$

where $I_{ph.j}$ is the function related to I_j and X_j for the jth receiver, at the same time, $I_{ph.j}$ also follow the same distribution model of X_j, and B_e is electrical filter bandwidth.

The outage probability is a commonly used performance metric in fading channels. It is defined as the probability that instantaneous SNR (γ_{total}) falls below a specified threshold SNR (γ_{th}), which represents a value of the SNR above which the quality of the channel is satisfactory. It is a useful method to evaluate the effect of the channel fading cause to the atmospheric turbulence on the system performance. The outage probability for a given threshold SNR (γ_{th}) can be defined as

$$P_{out}(\gamma_{th}) = P[\gamma_{total} < \gamma_{th}] \qquad (27)$$

For the case of MRC, the outage probability is given by [13]

$$P_{out}(\gamma_{th}) = \int_0^{\gamma_{th}} \int_0^{\gamma_{th}-\gamma_2} f(\gamma_1, \gamma_2) d\gamma_1 d\gamma_2 \qquad (28)$$

where $f(\gamma_1, \gamma_2)$ is the function of SNR follow the correlated Log-normal distribution.

5 Numerical Results

In this section, we present numerical results for the transmission performance of OFDM-FSO system with dual diversity reception under correlated Log-normal distribution. The main simulation parameters used in the numerical calculation are shown in Table 1.

Since the character of the channel correlation coefficient has been investigate in several works [6, 13], we prefer to recall the general points that will be useful while analyze the transmission performance to follow. Figure 3 shows typical curves for the channel correlation coefficient ρ_{12} as a function of separation distance under variation turbulence strength regimes represented by σ_R^2, and aperture size is given a fixed value $D = 2cm$. In this analysis, the optical wavelength $\lambda = 1550nm$ and transmission distance is 2km. From the graph, we can clear see that the channel correlation ρ_{12} decreases with increasing separation distance d_{12}, and the curves go cross a zero point where the channel correlation length ρ_c is defined (i.e. $\rho_{12} = 0$) under the variation turbulence strength σ_R^2 with fixed aperture diameter D. In general, the channel

Table 1 Numerical Parameters

Operating wavelength λ	1550nm
Relative intensity noise *RIN*	−130dB/Hz
PD load resistor R_L	50 Ω
Absolute temperature T_{abs}	300k
Transmission distance L	2km
Electron charge	1.602×10^{19}
Noise figure F_e	2dB
Electrical filter bandwidth B_e	2GhHz
Number of carrier N_s	256
Optical modulation index m_n	0.01
Detector responsivity R_D	0.8A/W

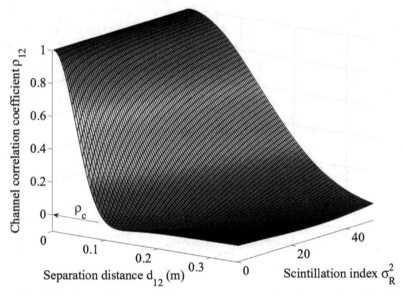

Figure 3 Channel correlation coefficient ρ_{12} versus separation distance d_{12} with variation of turbulence strength regimes

correlation length ρ_c is an important parameter which is useful in the choice of antenna spacings or making a quick judgment on whether channel correlations exist among component antennas. The obtained result shows the when the two channels are overlapped, they must be perfectly correlated. In fact, two separated channels are independent of each other when the separation distance is larger than channel correlation length ρ_c, and channel correlation is affected by turbulence strength. In addition, we consider the aperture averaging (AA) based on the previous literatures [14–16], the point receiver is defined as

$D = \sqrt{L\lambda}$, [4, 11]. Thus, the point receiver aperture size D is approximate to 2cm in this paper. The effect of AA on the system performance will be illustrated in Figure 6.

In Figure 4, we compare the system transmission performance of EGC and MRC combining scheme for OFDM-FSO link with different channel correlation coefficient, $\rho_{12} = (0.9, 0.1)$. It is observed that performance of EGC receiver is very close to MRC receiver when system under same channel correlation coefficient. The results also shown that there is a 2dB difference at average about BER $= 10^{-6}$ for system with $\rho_{12} = 0.1$ and $\rho_{12} = 0.9$, respectively.

In this case, receiver aperture size is such as aperture averaging, $D = 4cm$. In uncorrelated channel ($\rho_{12} = 0$), each aperture deployment far apart, and separation distance beyond to the channel correlation length ρ_c, the transmission performance can be seen as an independent case. This point of view is also in contract we typically see in RF wireless communication [13, 14], where only the uncorrelated channel has a lower bit-error-ratio, channel fading can be reduced substantially. Our observations demonstrate

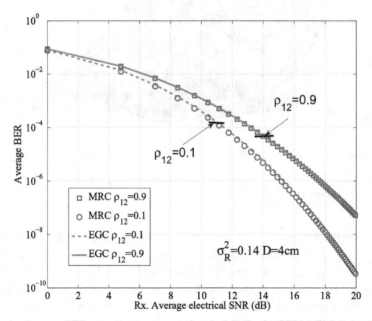

Figure 4   Average BER versus aver average electrical SNR of a OFDM-FSO link with dual diversity reception and MRC/EGC combining scheme over correlated Log-normal fading channel

that efficient separation between the apertures is crucial to achieve the promised improvement of system from independent dual diversity reception (channel correlation coefficient ρ_{12} approximate to zero, i.e. $\rho_{12} = 0.1$).

In Figure 5, we plot the derived average BER versus the average electrical SNR for different types of modulation, i.e., BPSK, QPSK, 16-QAM, 64-QAM, and adopt MRC combining scheme with channel correlation coefficient $\rho_{12} = 0.1$ as a independent case and $\rho_{12} = 0.9$ over weak turbulence channel, $\sigma_R^2 = 0.14$ with the point receiver aperture size $D = 2cm$. The performance of the proposed system is degraded when the correlation coefficient is high. However, for the average BER=10^{-10}, the BPSK with correlation co-efficient $\rho_{12}= 0.1$ outperforms BPSK with correlation coefficient $\rho_{12} = 0.9$ by approximately 5dB. These results indicate the importance of channel correlation and also demonstrate the effect of correlation on system performance. Moreover, the effect of modulation on the average BER in clearly apparent. When the average SNR = 14dB, for example, the average BER increases from 10^{-9} to $10^{-1.5}$ for QPSK and 64-QAM with same correlation coefficient $\rho_{12} = 0.1$.

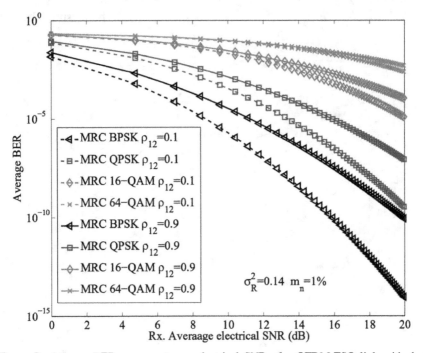

Figure 5 Average BER versus average electrical SNR of a OFDM-FSO link with dual diversity reception over correlated Log-normal fading channels

In fact, the average BER performance of dual diversity reception OFDM-FSO system with larger constellation size requires higher received power to accurately discriminate among the transmitted symbols; they are also sensitive to the turbulence due to the required precision in accurate constellation scaling at receiver.

To highlight the performance enhancement thanks to use of diversity receivers, we have shown in Figure 6 the system performance with ternary, dual and single reception in weak turbulence regime $\sigma_R^2 = 0.14$, and using MRC scheme for two different channel correlation coefficient $\rho_{12} = (0, 0.9)$. In this analysis, the effect of aperture averaging on system performance have been considered, i.e. $D = 4cm$ for single reception. It should be noted that the collected area of single receiver aperture is equal to the sum of the diversity apertures area in this case. Before presenting the numerical results, we defined the correlation Log-normal distribution function (22) for the case of ternary reception model. Based on the equation (13), (23) and (25), the electrical SNR at the output of diversity combiner as $\gamma_{total.MRC}(I_1, I_2, I_3, X_1^2, X_2^2, X_3^2) = \sum_{j=1}^{J} \gamma_j(X_j^2)$, $(j = 3)$, and average BER function as $\langle BER_{total} \rangle = \int_0^\infty \int_0^\infty \int_0^\infty f(X_1, X_2, X_3) BER_{total}(X_1^2, X_2^2, X_3^2) dX_1 dX_2 dX_3$. The result clear

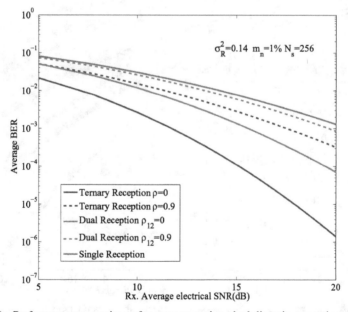

Figure 6 Performance comparison of ternary reception, dual diversity reception and single reception for OFDM-FSO link

show that ternary reception system performance outperforms dual and single reception case. For the received electrical SNR=15dB, for example, the average BER increases from 10^{-4} for ternary reception, 10^{-3} for dual reception to 10^{-2} for single reception with AA under same channel correlation coefficient $\rho_{12} = 0$. These result shown that selecting an diversity reception may increase the overall system performance and diversity technique can be obtained in practice through aperture averaging effect. Furthermore, we observe that there is a 5dB difference at average BER = 10^{-4} for the ternary reception with channel correlation parameter $\rho_{12} = 0$ and $\rho_{12} = 0.9$, respectively. As pointed out above, the result indicate that increasing the channel correlation leads to system performance degradation.

The system's outage probability P_{out} is analyzed using Equation(28). Figure 7 shows P_{out} in terms of the threshold SNR_{th} for single and dual reception with QPSK in the weak turbulence regime with the scintillation index $\sigma_R^2 = 0.14$ and optical modulation index $m_n = 0.01$. Two cases of channel correlation coefficient have been considered, $\rho_{12} = (0.1, 0.9)$. It is observed that the use of a dual reception leads to better P_{out}, which outlines its conspicuous

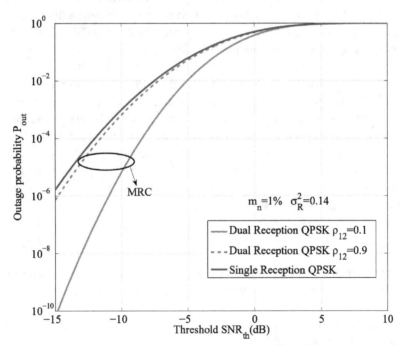

Figure 7   Performance of dual diversity reception with MRC and EGC for outage probability

contribution to mitigating the turbulence-induced fading overcoming the performance degradation. For instance, for a $SNR_{th} = -15$dB, P_{out} decreases from about 10^{-10} to 10^{-6} for a dual reception with $\rho_{12} = 0.1$ and single reception, respectively. Now, we consider the case of the channel correlation, $\rho_{12} = (0.1, 0.9)$ at the same outage probability $P_{out} = 10^{-6}$, we obtained a 4dB different for threshold SNR_{th} between $\rho_{12} = 0.1$ and $\rho_{12} = 0.9$ with a dual reception, respectively. The proposed system performance with channel correlation increasing leads to the performance degradation as the case of two channels are overlapped.

6 Conclusion

In this paper, we give a theoretical study of the channel correlation in the diversity reception OFDM-FSO system. In our analysis, we propose a novel transceiver architecture for atmospheric OFDM-FSO system with diversity reception over correlated fading channel. Based on the plane-wave model, we analyze and evaluate the channel correlation effects on the transmission performance of system over correlated Log-normal turbulence channel in terms of SNR, BER, outage probability and diversity combining scheme. Our numerical results demonstrate that diversity reception OFDM-FSO system performance is sensitive to the channel correlation, modulation format, aperture size and turbulence strength. The analysis results also represent spatial diversity as a helpful method to mitigate the channel fading and reduce the bit error of transmission. Furthermore, the evaluation of the proposed system behaviors outlines that the use of the diversity reception and OFDM technique can lead to substantial performance improvement, especially as high demand on transmission capacity becomes more important in next generation access networks. In the same time, our presented work can serve as preliminary guidelines to standard transceiver design for FSO communication system transmitting RF signals in the practical issue.

References

[1] H. Al-Raweshidy and S.Komaki, Eds. in *Radio Over Fiber Technologies for Mobile Communications Networks,* 1st ed. Artech House Publishers, 2002.

[2] *Co-location longitudinally compatible interfaces for free space optical systems,* ITU-T Rec. G.640, 2006.

[3] A.Bekkali. Transmission Analysis of OFDM-Based Wireless Services Over Turbulent Radio-on-FSO Links Modeled by GammaCGamma Distribution. *IEEE photonics journal*, 2(3):510–520, 2010.

[4] J.Armstrong. OFDM for optical communications. *J.Lightwave Technol.*, 27(3):189–204, 2009.

[5] L. C. Andrews and R. L. Phillips. in *Laser Beam Propagation Through Random Media*, Wellingham, WA: SPIE, 2005.

[6] Z.X. Chen. Channel correlation in aperture receiver diversity systems for free-space optical communication *Journal of Optics*, 2010.

[7] A.Bekkali, P.T.Dat, K.Kazaura, K. Wakamori, M.Matsumoto, T. Higashino, K. T-sukamoto and S.Komaki. Performance Evaluation of an Advanced DWDM RoFSO System for Transmitting Multiple RF Signals. *IEICE TRANS. FUNDAMENTALS*, 2009.

[8] S.Mohammad Navidpour. BER Performance of Free-Space Optical Transmission with Spatial Diversity *IEEE Transactions on wireless communications*, 26(8), 2007.

[9] X.Zhu and J. M. Kahn. Free-space optical communication through atmospheric turbulence channels. *IEEE Trans. Commun.*, 50(8):1293–1300, 2002.

[10] D.Skraparlis. On the Effect of Correlation on the Performance of Dual Diversity Receivers in Lognormal Fading *IEEE Communications Letters*, 14(11), 2010.

[11] M. A. Khalighi. Fading Reduction by Aperture Averaging and Spatial Diversity in Optical Wireless Systems. *J. OPT. COMMUN. NETW*, 1(2), 2009.

[12] Z.X.Chen, S. Yu, T.Y. Wang, G.H. Wu and S.L.Wang Channel Correlation in Aperture Receiver Diversity Systems for Free-Space Optical Communication *Journal of Optics*, 14(2012), 125710(7pp), 2012.

[13] F.Bai, Y. W.Su, T.Sato Performance evaluation of a dual diversity reception base on OFDM RoFSO systems over correlated log-normal fading channel *ITU Kaleidoscope Academic Conference*, 263–268, 2014.

[14] S.Bloom. in *The Physics of Free-Space Optics*, AirFiber Inc., 2002.

[15] L. C. Andrews Aperture-averaging factor for optical scintillations of plane and spherical waves in the atmosphere *JOSA A*, 14(2012), 9(4), 597–600, 1992.

[16] J. A. Anguita, M. A. Neifeld and B. V. Vasic Spatial correlation and irradiance statistics in a multiple-beam terrestrial free-space optical communication link *Applied Optics*, 46(26), 6561–6571, 2007.

[17] K. Kiasaleh Scintillation index of a multiwavelength beam in turbulent atmosphere *JOSA A,* 21(8), 1452–1454, 2004.

Biographies

Fan Bai received his B.E. degree in communication engineering from Changchun University of Science and Technology, Jilin Province, China, in 2007. From 2007 to 2010, he was with Beijing research and development center, ZTE Corporation, China, as system engineer. He received his M.Sc. degree in computer system and network engineering from the Waseda University, Japan, in 2013. He is currently pursuing his Ph.D. degree in Graduate School of Global Information and Telecommunication Studies (GITS), Waseda University, Japan. His research interests include optical wireless communications, optical fiber communications and digital signal processing. He is a student member of IEICE.

Yuwei Su was born in 1989 and received the B.E. degree from University of Electronic Science and Technoloy of China, Chengdu, China, in 2012.

He received the M.S. Degree from Waseda University, Japan. He is currently working toward the Ph.D. degree in the School of Fundamental Science and Engineering, Waseda University. His areas of interests include wireless communication, free space optics and optical science and technology.

Takuro Sato, professor. He received the Ph.D dgree in electronics from Niigata University, Japan, in 1994. From 1973 to 1995, he was with research and development laboratories, OKI electric Co., Ltd, Japan, as project leader. From 1995 to 2004, he was with department of information and electronics engineering, Niigata institute of technology, Japan, as professor. From 2004 to now, he was with graduate school of globe information and telecommunication studies as professor, Waseda university, Japan. His research interests include wireless communication, ICN/CCN, smart grid. Prof. Sato is a IEEE fellow, IEICE fellow.

Detecting and Mitigating Repaying Attack in Expressive Internet Architecture (XIA)

Beny Nugraha[1], Rahamatullah Khondoker[2], Ronald Marx[2], and Kpatcha Bayarou[2]

[1]*Department of Electrical Engineering, Mercu Buana University, Jakarta, Indonesia*
[2] *Fraunhofer SIT, Rheinstr. 75, Darmstadt, Germany*

Received: October 7, 2014;
Accepted: November 10, 2014

Abstract

Several Future Internet (FI) architectures have been proposed to address the problems of the Internet including flexibility (so called IP bottleneck), host-based addressing (addressing a host rather than the content itself), and security. In the beginning of this article, we survey the security solutions of seven FI architectures, namely XIA, RINA, NENA, SONATE, Mobility-First, NDN, and SONATE, based on literatures, prototypes, and demonstrations. It has been found that none of the architectures can fulfill all of the security goals: confidentiality, authentication, integrity and availability. Further in this article, we focus on eXpressive Internet Architecture (XIA) as it is the most secure and open-source Content-Centric Network (CCN). CCN is claimed by the Future Content Networks (FCN) Group to be the Future Internet. However, XIA does not have any mechanisms to mitigate the replaying attack, thus, this article proposes and implements a solution to mitigate it. Several existing solutions have been analyzed to derive the requirements for the proposed solution. By implementing the proposed protocol, XIA is now able to mitigate all of the reviewed network attacks. The evaluation shows that the proposed solution is more secure and less complex over the existing solutions.

Journal of ICT, Vol. 2, 151–186.
doi: 10.13052/jicts2245-800X.225

Keywords: Replaying Attack, Session Key, eXpressive Internet Architecture (XIA), Future Internet (FI), CCN.

1 Introduction

Current Internet faces challenges such as inability to provide flexibility - changing of protocol in one layer requires another changing of protocol in another layer, and inability to provide intrinsic security - a security mechanism is added to counter a new threat, it is not integrated. The problems arise mainly because of the design principles of the Internet that are hard to be changed (cannot provide flexibility) [1]. Several Future Internet (FI) architectures have been developed to solve these problems. There are two design methods that can be followed for developing an FI Architecture: "clean slate" or "evolutionary". In the clean slate approach, the architecture is designed from the scratch, meanwhile new design components are added to the existing architecture in the evolutionary approach. In the early part of this article, we analyze the security mechanisms of seven future network architectures, namely, eXpressive Internet Architecture (XIA) [2], Recursive Inter-Network Architecture (RINA) [3], Service Oriented Network Architecture (SONATE) [4], Netlet-based Node Architecture (NENA) [5], MobilityFirst [6], NEBULA [7], and Named Data Networking (NDN) [8]. We selected them as they are mature (established in 2009 or 2010), and they have either a demonstration or a prototype or both. NENA and SONATE both use clean slate approach. Meanwhile the other five, XIA, RINA, MobilityFirst, NDN, and NEBULA, all of them use evolutionary approach.

It is indispensable for a newly deployed Internet architecture to fulfil the security requirements. In this article, we discuss the following security goals: confidentiality, integrity, availability, and authentication (defined in Section 2.1). The methodology of the research are, first, specify the threats against each of the security goal (discussed in Section 2.2), second, identify the available security mechanisms of each architecture by analyzing its literatures, prototype, and demonstration (described in Section 2.5) and third, conclude which of the threats can be mitigated by which of the security mechanisms (depicted in Section 2.5).

The rest of this article is organized as follows: the result of the analysis of the FI's security mechanisms is presented and compared in Section 2 and Section 3, respectively. Afterwards, in Section 4, the existing solutions for replaying attack are analyzed to derive the requirements for the proposed solution which is described in Section 5. Based on the derived requirements, the solution is proposed in Section 6. The implementation and evaluation of the proposed solution are described in Section 7 and 8 respectively. Finally, the conclusion and the future work are discussed in Section 9.

2 Methodology for the Survey

The security mechanisms of an FI can be analyzed in one of the two following ways: attack-centric and system-centric [9]. Attack-centric means the attacks on a system (i.e., using attack trees) are modeled, and system-centric means the system itself (i.e., using STRIDE methodology) is modeled. Our methodology is a combination of attack-centric and system-centric approach since it analyzes both the architectures and the attacks, and it is able to provide a better view of the architectures vulnerability to attacks than to follow just one approach. Our methodology works as follows:

1. Defining the network security goals.
2. Specifying attacks that inhibit the network security goals and then analyzing the counter mechanisms for each attack. This is the attack-centric approach.
3. Selection of future network architectures.
4. Analyzing the security solutions of each of the architectures. This the system-centric approach.
5. Matching the counter mechanisms for each attack and the security mechanisms of the future network architectures in order to find out the vulnerability in each architecture.

The details of each item will be discussed as follows:

2.1 Defining Security Goals

The general goal of network security is to give people freedom to enjoy computer networks without fear of compromising their rights and interests [10].

In order to achieve that goal, four specialized goals of network security have been identified. These four goals are the following:

Confidentiality Means that the message that is sent by the sender has to be intended for the receiver only, for the others, this message must be worthless.

Integrity Means that the received message must be the same as the original message.

Availability The services that are accessible by the Internet (i.e., web services, remote machines, networks, etc.) must be available all the time for its authorized users only.

Authentication Only the authorized user is able to send a message and the receiver is able to proof the sender's identity.

2.2 Specifying Threats Against Achieving the Goal

Several network security threats which work against achieving the goal have been identified. They are as follows:

2.2.1 Threats against confidentiality

Snooping and traffic analysis attacks are considered as possible threats against confidentiality. In snooping, the aim of the attacker is to get the database of an authorized user or the packets flowing in a network. The attacker can perform several action to undergo snooping attack, examples of the action are: 1, by using ping-type programs (ICMP ping, TCP ping) to identify active hosts on the network and to further locate potential targets and, 2, by using TCP/UDP port scanning for detecting the target operating system [11]. Snooping attack can be mitigated by having a data encryption mechanism to protect the packets.

In order to perform the traffic analysis attack, the attacker intercepts and examines messages to extract information from the traffic patterns in a communication. The greater the number of packets that can be obtained, the more the information that can be inferred from the traffic. The security mechanism to prevent this attack is to have a mechanism to conceal the identity of the users, therefore, an attacker cannot determine at which point or node he should watch the traffic.

2.2.2 Threats against integrity

Modification and repudiation attacks are threats against integrity.

Modification attack involves deletion, insertion, or alteration of information in an unauthorized manner that is intended to appear genuine to the user [12]. The counter mechanisms for this attack are to hash the message or to have a digital signature, therefore, the receiver will be able check the correctness of the message.

Repudiation is a process in which the sender or the receiver cannot prove that a transaction has taken place between them, either one or both of them can deny that they are sending or receiving the data [12]. Repudiation attack can be mitigated by having a digital signature mechanism in collaboration with a trusted third party to create a non-repudiation message.

2.2.3 Threats against confidentiality

Denial of Service (DoS) attack is a threat against availability. This attack can deny access to information, applications, systems or communications. An example of DoS attack is to flood the traffic with bursts of packets [13]. DoS can be prevented by having a flow control or bandwidth allocation mechanism. Therefore, only the authorized packets that can flow in the traffic.

2.2.4 Threats against authentication

The attacks that are against authentication are: man-in-the-middle, reflection, masquerading, and replaying attack.

The attacker stays in between the sender and the receiver, then observes or modifies the traffic in the man-in-the-middle attack [14]. Man-in-the-middle attack can be prevented by having a digital signature mechanism in order to authorize the real authorized users.

In reflection attack, the attacker has an objective to pretend that he is an authorized user by sending the response from the real authorized user to the target [15]. Reflection attack can also be mitigated by performing a digital signature mechanism.

In masquerading, the attacker pretends to be an authorized user of a system in order to gain access to it, and then modifies the [16]. Masquerading attack can be mitigated by having an anonymous connection or having a good user authentication process.

Replaying attack occurs when information is captured and then replayed later, in different session, in order, for example, to gain the trust of other users [17]. Replaying attack can be prevented by having a marker to bind one communication session, example of the marker are session key or random number which will be generated differently each session. By performing this, the messages in one session will be always different than the messages in another session.

To summarize, there are nine attacks to be reviewed in this article. The future network architectures should be able to mitigate all of the attacks intrinsically to fulfill the security requirements.

2.3 Selection of FI Architectures

The selection of future network architectures is done by considering the maturity of the architecture (architectures that established from year 2011 onward will not be considered), the availability of the demonstration, and the prototype as shown in Figure 1. We choose seven architectures to review, they are eXpressive Internet Architecture (XIA), Recursive Inter-Network

Criteria	Future Internet Architectures						
	XIA	RINA	SONATE	NENA	MobilityFirst	NEBULA	NDN
Approach	Content-Centric	Content-Centric	Protocol Graph	Protocol Graph	Content-Centric	Supports Cloud Computing	Content-Centric
Project Started	In 2010	In 2010	In 2009	In 2009	In 2010	In 2010	In 2010
Demo	√	√	√	√	√	√	X
Prototype	√	√	√	√	√	X	√

Legend:
√: Available
X: Not Available

Figure 1   FI Architectural approaches

Architecture (RINA), Service Oriented Network Architecture (SONATE), Netlet-based Node Architecture (NENA), MobilityFirst, NEBULA, and Named Data Networking (NDN) as they are mature (the projects started in 2009 or 2010), and they have either a prototype or a demonstration or both.

2.4 Asking and Receiving Expert's Feedback

After surveying the literatures for each architecture, we got the result described in the following section 2.5. The result of each architecture has been sent to the founders of the architecture for review such as we asked the feedback for the XIA architecture only from the XIA inventors. We received feedback for all of the seven architectures, this step is needed to check whether our analysis is correct or not.

2.5 Analyze the Security Solutions of Every Architecture

The first two architectures that are analysed are SONATE and NENA. The aim of both SONATE and NENA is to use a customized protocol graph (similar to a TCP/IP or UDP/IP network stack) based on the requirements from the application. However, they differ in terms of "when the composition is accomplished?". Whereas in SONATE, the composition is done during runtime of communication association, the composition in NENA is accomplished during the design time of creating new protocols.

2.5.1 SONATE

In SONATE, the services provided by building blocks (the implementation of a protocol or a mechanism like CRC, retransmission, etc.) are selected and composed by a composition algorithm to create a protocol graph during runtime based on the requirements from the application, constraints from the administrator, and networks [4]. Therefore, the security mechanisms in SONATE are depends on the application's requirements. The security mechanisms are [19]:

1. SONATE is able to select building blocks (BB) that:
   a. enable data encryption (i.e., data encryption micro protocol) [19],
   b. provide data authentication (i.e., digital signature, MAC) [19,20], and
   c. provide flow control service [20].
2. Each communication session is bound by one protocol graph [19].

By analyzing the above security mechanisms, we can conclude that SONATE is vulnerable to the traffic analysis attack and the masquerading attack since SONATE does not provide anonymous communication and an attacker only need to know the port and the address of the target to initiate both attacks. SONATE also cannot mitigate the repudiation attack because it does not have a trusted third party to prove a communication between two users has been finished, thus, cannot create a non-repudiation message.

The advantage of SONATE is that it is able to mitigate the other attacks by selecting an appropriate BB to counter the attacks. Example of BB that can be used for counter the threats are data encryption (can be used to prevent the snooping attack), digital signature (can be used to mitigate the modification, man-in-the-middle, and reflection attacks), and flow control (can be used to counter DoS attack).

2.5.2 NENA

In NENA [5], the services provided by building blocks (the implementation of a protocol or a mechanism like CRC, Retransmission, etc.) are selected and composed by a composition algorithm to create a protocol graph called netlet. The composition process run during design time (by a developer or assisted by a software) assuming the requirements from an application, constraints from the administrator and networks. However, the selection of the most appropriate netlet is accomplished during runtime.

The security mechanisms of NENA are:

1. NENA uses secure deployment of protocols. Each protocol has a unique protocol ID [21].
2. NENA has a collaborative attack detection mechanism [22].
3. Similar to SONATE, NENA is able to select a protocol that offers:

    a. Data encryption (i.e., data encryption netlet) [5].

    b. Data authentication (i.e., digital signature, MAC) [5].

By matching the above security mechanisms with the attacks, we conclude that NENA is not able to prevent the traffic analysis and masquerading attacks since NENA does not provide anonymous connection and an attacker only need to know the port and the address of the target to pretend to be the target. Furthermore, NENA vulnerable to the repudiation attack since it does not have a trusted third party server.

However, NENA is able to mitigate the other attacks by using the secure deployments of protocols, collaborative attack detection, or by selecting an appropriate netlet to counter the attacks. Examples of netlets that can be used for counter the attacks are data encryption and digital signature netlets. Meanwhile, in order to mitigate denial-of-service attack, NENA utilizes the service of collaborative attack detection.

2.5.3 XIA

Whereas in the Internet, an IP address is used to address both the host and the content, XIA uses three principle types of identifiers to retrieve the content: Content ID, Host ID, and Service ID [2]. The content ID, which is the hash of the content, is used to retrieve the content without needing to know its location. The host ID, which is the hash of the host's public key, is used to contact the host that provides the content. The service ID, which is the hash of the service's public key, is used to get the service that provides the content. The security mechanisms in XIA are:

1. The architecture uses Content/Host/Service ID (CID/HID/SID) in order to retrieve the content. CID is the hash of the content, HID is the hash of the Public Key of the Host, and SID is the hash of the Public Key of the service [2].
2. XIA has the LAP (Lightweight Anonymity and Privacy) defence mechanism enables anonymous communication to prevent remote tracking [23].
3. The STRIDE defence mechanism allocates the available bandwidth in a tree-based topology [24]. This mechanism also available in XIA.
4. XIA has the AKI (Accountable Key Infrastructure) defence mechanism which provides a reliable data authentication process [25].

By reviewing the security mechanisms in XIA, we conclude that XIA is able to mitigate all of the reviewed attacks except the replaying attack, because XIA does not have any mechanism to bind one communication session.

For snooping attack, XIA mitigates them by using the public key and private key of a service (SID) to do encryption mechanism. For the other attacks, XIA mitigates them by using a hashed ID (CID/HID/SID) or by using the defence mechanisms provided by SCION architecture (LAP, AKI, or STRIDE defence mechanisms).

2.5.4 RINA

The basic design principle of RINA is that "Networking is only Inter-Process Communication (IPC)" [3] [26]. IPC is a function to allow two communication processes (one in the sender and another in the receiver side) to communicate with each other. Examples of the IPC functions are: locating processes, determining permissions, passing information, scheduling, and managing memory. Process names are used as identifiers. For example, a source application process requests a service using the process name of the destination application. They communicate with each other by utilizing the services of the Distributed IPC Facility (DIF).

The security mechanisms in RINA are:

1. All members in the same DIF must be authenticated first before they can join in [27].
2. Even if the attacker is already inside the DIF, he still needs to scan all of the possible Connection End Point id (CEP-id) of the target, and the probability is 2^{16} (given that the CEP-id is 16-bit) [27].
3. RINA has a SDU protection module that is able to provide security functions such as: encryption function, compression function, and error detection function [28].
4. The CEP-ids in RINA are used to distinguish between the new and the old data connection [27].

By analyzing the above security mechanisms, we acknowledge that RINA is able to mitigate all of the reviewed attacks except the DoS and repudiation attacks. The research to prevent DoS attack from the inside is on going since it is hard to detect. Meanwhile, RINA does not have a trusted third party to prevent the repudiation attack.

For the other attacks, RINA mitigates them by utilizing the function of the SDU Protection Module, the authentication process by IPC Management, and the unique CEP-Id that is assigned to each user. The CEP-Id can also be used to distinguish the old and new communications. The CEP-Id can mitigate the replaying attack by distinguishing the messages from a new session and the messages from an old session.

2.5.5 MobilityFirst

In this architecture, the end-user can request for a service using the Human-Readable Name (HRN). The naming architecture of the MobilityFirst has three identifiers: Network Address (NA), Globally Unique Identifier (GUID), and HRN [5]. It ensures mobility by separating network location information so called NA from its identity so called GUID. Similar with XIA, GUID is the hash of the content itself. MobilityFirst has two mapping services: Name Assignment Services (NAS) and Global Name Resolution Service (GNRS). The NAS binds an HRN with GUID and the GNRS maps GUID to NA. GNRS functions as a content location directory as it dynamically binds the name and the location. When the content is available in more than one locations, GNRS chooses the content for the requester from the nearest location.

The security mechanisms in MobilityFirst are:

1. To retrieve the content, MobilityFirst uses Globally Unique Identifier (GUID) which is assigned to each content as an address [6]. GUID is a result of hashing the content and can be used as a public key for encryption mechanism.
2. MobilityFirst enables frequent routing update using the function of GNRS [29].
3. MobilityFirst uses an integrated protocol that enables self-certifying public key names [30].

By matching the security mechanisms with the attacks, we conclude that MobilityFirst is robust against the snooping attack because it has a mechanism to encrypt the packets by using the name or GUID of the content as the identifier. For the modification attack, MobilityFirst mitigates it by assigning a unique GUID to each content. GUID is a hash of a content, therefore, the receiver can check the correctness of the content. For the man-in-the-middle, reflection, and masquerading attacks, MobilityFirst mitigates them by using a protocol that is able to authenticate a user and having a unique GUID for every content. For the DoS attack, MobilityFirst prevents it by utilizing the function of GNRS to perform routing update. For the repudiation attack, MobilityFirst prevents it by having a non-repudiation message, which is created by the PKI.

However, MobilityFirst cannot mitigate the traffic analysis and replaying attacks due to the following reasons:

1. MobilityFirst cannot mitigate the traffic analysis attack because it does not have a mechanism to enable anonymous communication.
2. The replaying attack is possible to do in MobilityFirst since it does not have a mechanism to bind the messages with the sessions.

2.5.6 NDN

NDN [8] defines two types of packets: one is for request (called interest packet) and another is for reply (called data packet). The interest packet has mainly two fields: content name and nonce (number once). The content name identifies the data to be retrieved and the nonce binds each communication session. The data packet carries both the name and the content of the data, together with the digital signature and signed information.

The security mechanisms in NDN [31] are:

1. An end-to-end encryption can be used in NDN. This is used to encrypt the NDN data.
2. Data packets are signed using the digital signature.
3. The clients in NDN send nonce (Number Once) within the interest packets.

By analyzing the above mechanisms, it can be concluded that NDN is vulnerable to the traffic analysis attack even though NDN is a content-centric network. Being content-centric is not enough to prevent that attack, it needs a mechanism such as a mechanism to conceal the packet, and this mechanism is not provided by NDN. The research for a method to mitigate the DoS attack is on going, therefore, NDN is vulnerable to the DoS attack.

NDN is able to mitigate the replaying attack since it has a nonce in its interest packet. This nonce will differentiate the old and the new interest packet. For the modification, repudiation, man-in-the-middle, and reflection attacks, NDN prevents them by performing a digital signature mechanism. For the masquerading attack, NDN prevents it by having a digital signature and unique name for every content.

2.5.7 NEBULA

NEBULA [7] facilitates data centers in a cloud environment to communicate in a reliable way. NEBULA consists of three components: NEBULA Core (NCore), NEBULA Data Plane (NDP), and NEBULA Virtual and Extensible Networking Techniques (NVENT). NCore interconnects the data centers using a reliable routing mechanism. NDP is a data plane that provides flexible access control and security mechanisms. NVENT is a control plane, which responsible for determining paths for the packets to arrive at the destination.

The security mechanisms in NEBULA [32] are:

1. Proof of Consent (PoC) mechanism to authorize a packet and a path.
2. Proof of Path (PoP) in order to make sure that the packet only flows on the authorized path.

3. NEBULA uses token to bind one authorized communication session.
4. There is a consent server in NEBULA that can act as a trusted third party. This server will prove that the communication between two users really took place because the users who want to send a packet will contact it to obtain the PoC.

NEBULA was not designed to mitigate the snooping and traffic analysis attacks. To mitigate these attacks, mechanisms such as end-to-end encryption or onion routing need to be applied on top of NEBULA. NEBULA also cannot prevent the masquerading attack because the attacker can pretend to be the authorized user by getting the users address.

NEBULA is able to mitigate the other attacks by using the function of a PoC and a PoP that reside in the NDP. Moreover, NEBULA is able to prevent repudiation attack because it has a consent server as a trusted third party to create a nonrepudiation message.

3 Survey Results

The result of the comparison of all architectures is shown in Figure 2. It can be seen that SONATE and NENA cannot mitigate the same attacks which are the traffic analysis, masquerading, and repudiation attacks. They cannot mitigate the traffic analysis attack because they do not have any mechanism to enable anonymous communication. SONATE and NENA cannot prevent the masquerading attack since they do not provide anonymous communication and the attacker can get the IP address of the communicating hosts to pretend to be the authorized users. Meanwhile, they cannot prevent the repudiation attack because they do not have a trusted third party.

In summary, none of the studied architectures, whether it is clean slate or evolutionary, can mitigate all of the reviewed attacks. However, it can be seen in Figure 2 that XIA is the most secure, since it is able to handle eight out of nine reviewed attacks. Moreover, XIA is Content-Centric Network (CCN). CCN is claimed by the Future Content Networks (FCN) Group as the Future Internet (FI) [17]. At last, XIA has both a demonstration and a prototype available on github [33], thus, XIA can be considered as the most promising architecture to be deployed into the market.

Furthermore, in this article we choose to focus more on XIA because of the above reasons. Since XIA is still vulnerable to replaying attack, in this article we provide a replaying attack solution which is more secure and less complex than the existing solutions. In order to provide that solution, first we analyze the existing solutions for replaying attack in order to derive the requirements

Security Goals	Security Attacks	SONATE	NENA	XIA	RINA	MobilityFirst	NDN	NEBULA
Confidentiality	Snooping	√	√	√	√	√	√	X
	Traffic Analysis	X	X	√	√	X	X	X
Integrity	Modification	√	√	√	√	√	√	√
	Repudiation	X	X	√	X	√	√	√
Availability	Denial of Service	√	√	√	X	√	X	√
Authentication	Man-In-The-Middle	√	√	√	√	√	√	√
	Reflection	√	√	√	√	√	√	√
	Masquerading	X	X	√	√	√	√	X
	Replaying	√	√	X	√	X	√	√

Legend:
√: Can be mitigated
X: Cannot be mitigated

Figure 2   Comparison of all architectures in terms of handling attacks

for the proposed solution. The analysis of the existing solutions is provided in the following Section.

4 Analysis of the Existing Solutions for Replaying Attack

In order to propose a solution to be implemented in XIA, we reviewed the advantages and disadvantages nine existing solutions that have the mechanisms to prevent the replaying attack (e.g., session keys, random nonce, or timestamp). They are Diffie-Hellmann [34], Lamport's Password Authentication [35], S/Key One Time Password [36], Keung-Siu Protocol [37], Message Binding [38], Timestamp [39], Luo-Shieh-Shien Authentication Protocol [40], Yoon-Jeon Protocol [41], and Tseng-Jou Protocol [42]. The review of each solution is presented in the following Subsections.

4.1 Diffie-Hellman

Diffie-Hellman is a method to compute a unique session key. In order to compute a session key, the sender and the receiver choose two public parameters and generate a new private value in every session. Diffie-Hellman was developed by Whitfield Diffie and Martin E. Hellman and it was published in 1976 [34].

Advantage: This method is considered to be secure if the value of the public parameters, p and g, are chosen properly. Therefore, it is not likely for an attacker to calculate the secret key s= gab mod p. The secret key can be used to prevent replaying attack because only with the correct secret key Alice and Bob can encrypt and decrypt their messages [34].

Disadvantage: Original Diffie-Hellman scheme does not authenticate the communicating users, thus, it is vulnerable to the man-in-the-middle attack [42]. A person in the middle may establish two distinct Diffie-Hellman key exchanges, one with Alice and the other with Bob, effectively masquerading as Alice to Bob, and vice versa, allowing the attacker to decrypt (and read or store) then re-encrypts the messages passed between them.

4.2 Lamport's Password Authentication

Lamport's password authentication is a secure one-time password authentication method that was published by Leslie Lamport in 1981 [35,43]. This method implements a one-time password to protect against eavesdropping. The authentication process is between the user (A) and the server (S).

Advantage: This method is robust against the replaying attack since one session is bound by one password. Furthermore, a system that uses this method will never use a same password even though the system is crash. The system does not require back up to a point where a password already have been used, the system will continue from the point when the system crashed.

Disadvantage: This method is vulnerable to one type of man-in-the-middle attack, called the small n attack (e.g., the attacker impersonates the server).

4.3 S/KEY One Time Password

S/KEY One Time Password is a method that only allows one password ever crosses the network. The secret of a user will never be shared, thus, it prevents from an eavesdropping. This method was published by Neil Haller in 1994 [36].

Advantage: The user's secret pass-phrase never crosses the network at any time, thus, this method is able to prevent an eavesdropper. Assuming that an attacker manages to get hold of a password that was used for a successful authentication. This password is already useless for subsequent authentications, because each password can only be used once in one session, thus, prevents replaying attack.

Disadvantage: This method is vulnerable to a dictionary attack where the attacker is using a list of most possibly used passwords to guess the secret [44].

4.4 Keung-Siu Protocol

This protocol was developed by Stephen Keung and Kai-Yeung Siu in 1995 [37]. Aims of this protocol are to establish a session key while protecting the weak passwords (easy to be guessed by using a list of commonly used passwords) and to prevent off-line password guessing attack (the attacker guesses a password by analyzing the pattern of legitimate user's password). This protocol provides authentication process by using challenge and response messages that allow both users to validate each other.

Advantage: The protocols are immune to the replaying attack because of the following properties:

1. It uses random number that is different in every session. This random number is used to ensure that both hosts are communicating in one session.
2. This method also uses session key that is different in every session. The session key can be used to prevent the replaying attack.
3. This method uses encryption mechanisms so that the attacker is unable to read the message.

Disadvantage: The source and the destination must know the public key server, meanwhile, there is a situation where the public key is difficult to obtain (e.g., in a mobile environment).

4.5 Message Binding

By binding the messages to their correct context (e.g., binding the message to its protocol run), the replaying attack can be prevented. One way of binding the message can be done by including an information in the messages, therefore they are recognized to belong to a certain state of a certain protocol run. Example of information that can be included in the message is a protocol identifier [38].

Advantage: Message binding is able to withstand replaying attack because it has an information that is tagged to the message to bind the message and the protocol run.

Disadvantage: Message binding cannot bind a message and a session, it only binds a message with a protocol run. That means, in a certain point the replaying attack cannot be prevented (e.g., where the same protocol is used in a different session).

4.6 Timestamp

Timestamp is a marker that is used in a message to ensure the freshness of the message [39].

Advantage: The replaying attack is prevented by the use of timestamps. For example, a developer sets the value of δ_{max}, a constant to limit the difference in timestamp, to 200 milliseconds. If the receiver gets the message and the value of $|T_{sender} - T_{receiver}|$ is higher than 200 milliseconds, then the receiver will detect the replaying attack and drops the message [39].

Disadvantage: One disadvantage of timestamp is in term of clock synchronization of the two hosts. Synchronization is required to maintain the accuracy and precision of the timestamp. The other disadvantage is, maintaining a list of used timestamps within the current window has the drawback of potentially large storage requirement, and corresponding verification overhead [41].

4.7 Luo-Shieh-Shien Authentication Protocol

This is a protocol to generate session keys with the help of a third party authentication server. This protocol was developed by Jia-Ning Luo, Shiuhpyng Shieh, and Ji-Chiang Shen and was published in 2006 [40].

Advantage: This protocol uses random numbers and session keys. The replaying attack can be mitigated by using the session keys as marker to distinguish the messages in different sessions. Furthermore, there is a mechanism to ensure that both hosts have created the same session key.

Disadvantage: There is a redundant message that increases the complexity of the protocol.

4.8 Yoon-Jeon Protocol

This protocol was developed by Eun-Jun Yoon and Il-Soo Jeon and was published in 2010 [41]. The protocol generates session keys based on Chebyshev polynomial.

Advantage: This protocol is robust against the replaying attack by utilizing the session key to ensure that the messages in one session are different than the messages in another session. Additionally, secure mutual authentication between entities is achieved by using a MAC by each entity. MAC is created by hashing the identity of both users and the Chebyshev Polynomial that is received by each user.

Disadvantage: There is an unused random number N. User A selects large prime number N that is not used in any operation and it is also not used to detect the freshness of the message. The inclusion of an unused random number can increase the complexity of the protocol.

4.9 Tseng-Jou Protocol

This protocol is an improvement of Yoon-Jeon Protocol. Similar to Yoon-Jeon Protocol, Tseng-Jou Protocol uses Chebyshev polynomial as a base to generate session keys. The main improvement is, it provides anonymous identity of the hosts by generating a parameter pseudo identity (PID) on each host. This protocol was developed by Huei-Ru Tseng and Emery Jou and published in 2011 [42].

Advantage: This protocol can mitigate replaying attack by using the session key to ensure the messages are bound to a specific session. It also provides anonymous identity of the host by having parameter PID on each host.

Disadvantage: There is an unused random number N_i. User U_i selects a large prime number Ni that is not used in any operation. To decrease the complexity of the protocol, the used of an unused random number can be avoided.

5 Derived Requirements for the Proposed Protocol

It can be seen in the last Section that all of the reviewed existing solutions have their own problems.

The properties that need to be satisfied by the proposed protocol are:

1. Use of a marker to distinguish the messages in different sessions.
2. Having a process to ensure that both users generate the same session key.
3. Using an encryption mechanism to protect the message, therefore, it is unreadable by the attacker.
4. Utilizing a mechanism to conceal the identity or the address of the sender.

Meanwhile, the properties that need to be avoided by the proposed protocol are:

1. Even though timestamp can be used as a marker, it has a disadvantage in term of clock synchronization between two communicating users. Therefore, timestamp can be avoided to reduce the risk of having synchronization issue.
2. Redundant computation that reduces the efficiency of the protocol.
3. The use of a useless random number that increases the complexity of the protocol.
4. To use several encryption mechanisms that reduces the efficiency of the protocol.

6 The Proposed Protocol

The proposed solution has to satisfy the desired properties and avoid the unwanted ones. Thus, the proposed solution is a complete protocol that provides a mechanism to mitigate replaying attack, provides an encryption mechanism, enables anonymous connection, and provides mutual authentication process. The protocol has the following properties:

1. It has markers in each session in the form of session keys (each host has one session key with length up to 280 bits).
2. The session keys are generated by XOR computation of four random numbers (70 hex per random number). The session keys are used by both users to differentiate the messages in different sessions.
3. Has a mechanism to ensure that the random numbers that are received at the receiver side are correct. This mechanism is needed for both hosts to create the same session key. This is achieved by checking the MAC in each host. The MAC value that is sent by User B has the random numbers that is generated by User A and has been received by User B. If User A finds the difference in the MAC value (e.g., someone is altering the random numbers, or there is an error in the network so that User B cannot obtain the random numbers from User A), then User A will terminate the session.
4. It also has a mechanism to ensure that both users generate a correct session key. This mechanism is needed to detect the replaying attack. This is also achieved by checking the MAC in each host and if each user has verified the MAC, then both users has generated a same session key.
5. Has two times data encryption, therefore, an attacker cannot read the message.
6. It does not have redundant computation and useless random number, thus reduces the complexity and increases the efficiency of the protocol.
7. It generates a parameter that is called Pseudo Identity (PID) to hide the host's identity.

The sequence diagrams of how the protocol works can be seen in Figures 3 and 4.

The proposed protocol works in the following ways:

1. The first assumption before running the protocol is as follows: User A and user B have exchanged their public key to be used for the encryption-decryption mechanisms.
2. User A generates two random numbers n_{A1} and n_{A2}, hashes these two random numbers, then computes pseudo identity (PID_A) to hide his identity. He encrypts his identity (HID_A) and his random numbers with the user B's public key, and then sends it along with the PID_A to user B.
3. User B generates two random numbers n_{B1} and n_{B2}, hashes these two random numbers, then computes pseudo identity (PID_B) to conceal his

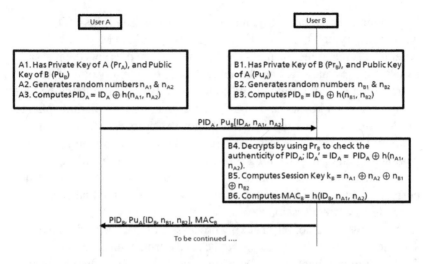

Figure 3   Sequence diagram of the proposed protocol-top part

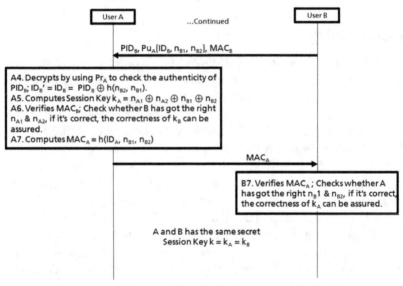

Figure 4   Sequence diagram of the proposed protocol-bottom part

identity. He encrypts his identity (HID_B) and his random numbers with user A's public key. After receiving the message from A, he decrypts the message using his private key, then he authenticates the identity of A, if it is not correct then he will terminate the connection. But if it is correct, he will compute session key k_B and MAC_B. Then he sends his PID_B along with the encrypted message (HID_B, n_{B1}, n_{B2}) and the MAC_B.

4. After user A receiving the message from user B, he decrypts the message by using his private key. Then he authenticates the identity of B, if it is not correct then he will terminate the connection. But if it is correct, he will compute session key k_A and MAC_A. After that, user A will authenticate the MAC_B to make sure that user B got the correct random numbers from him, this means B also has generated a correct session key. After completing all of the checking processes, user A sends his MAC_A to B.

5. User B will authenticate MAC_A to make sure that user A has got the correct random numbers from him, and also to make sure that user A has generated a correct session key.

6. After completing all of the checking processes, user A and user B have the same secret session key ($k_A = k_B$) to be used during their communication.

7. The random numbers and the session keys that are generated by user A and user B are different in every session.

7 Implementation

The protocol is implemented in XIA Prototype in order to prove that the protocol is able to make XIA robust against replaying attack and is able to generate the desired result (secured session keys). In order to simulate how the proposed protocol prevents the replaying attack, a topology is created by using VirtualBox version 4.2.12. The topology can be seen in Figure 5.

It can be seen in Figure 5 that the Attacker is connected to the Router via Ethernet 1 and to the NAT via Ethernet 2. In XIA, The Attacker cannot connect to Host0 and Host1 directly. It is necessary for the Attacker to connect with the Router. Since the HID of the Attacker is given by the Router. The Router is the one that connects the Attacker with Host0 and Host1. Also can be seen in Figure 5 that each host connected via two interfaces, one of them is connected to the Router while the other is connected to NAT. Every hosts need to be connected to NAT in order to give them an internet connection that is used to obtain the XIA Prototype from the github [33].

There are three cases to be used to test the proposed protocol: First, a case when Host0 and Host1 are sending and receiving data without being interrupted by the Attacker. In this case, Host0 and Host1 do not run the proposed protocol applications. Second, when the Attacker is successfully performing the replaying attack. In this case, Host1 authenticates the Attacker as Host0. Third, a case when Host0 and Host1 run the applications for the

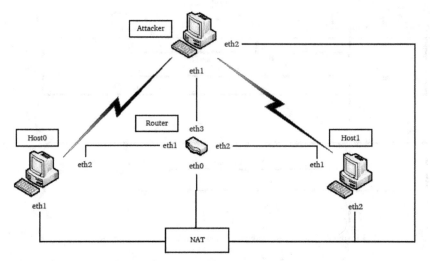

Figure 5   Topology for implementation

proposed protocol before they start exchanging data. This case is used as a proof that the protocol is able to mitigate the replaying attack.

7.1 Common Data Exchange

In this case, Host0 and Host1 are exchanging data without using the proposed protocol. Instead of using the session key as a header of a data, they use their HID. To use the HID as a header, they are exchanging their HID before they start exchanging data. The sequence diagram of this step can be seen in Figure 6. It can be seen that, Host0 and Host1 communicate in one session only. It is assumed that the Attacker is idle.

The result from this case is, each host uses its HID as a header of the message that it wants to send. The HID is used by the receiver to authenticate the sender.

7.2 Replaying Attack Scenario

This case is to simulate the replaying attack. It is assumed that Host0 and Host1 have already exchanged HID. These HIDs are always same in each session. The scenario is, the Attacker captured and saved the data from Host0. To simulate the replaying attack, it is assumed that the previous session has ended and the Attacker replays the data from Host0 to Host1 in the next session. The sequence diagram of this step can be seen in Figure 7.

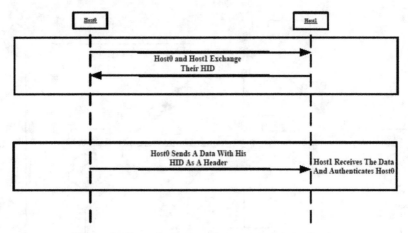

Figure 6   Scenario diagram for common data exchange

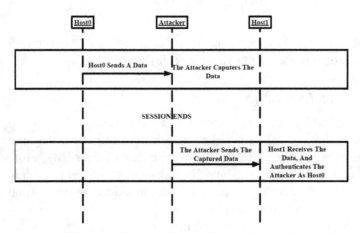

Figure 7   Sequence diagram for replaying attack scenario

The result from this case can be seen in Figure 8 to 11.

Figure 8 represents the following processes:

- Host0 used his HID, "This Is The Real Host0", as the header of the data. The data consists of the HID of Host0 and the message that he wants to send, "Test Replaying Attack".
- Host0 sent the data, and this data is intended to be sent to Host1.

It can be seen in Figure 9 that the Attacker got the data from Host0 and then save it.

```
beny@beny-VirtualBox:~/xia-core/bin$  ./dataExchangeServer
Stream service started

Stream DAG
DAG 0 -
AD:1000000000000000000000000000000000000000 1 -
HID:0000000000000000000000000000000000000000 2 -
SID:0f00000000000000000000000000000000000888

registered name = www_s.replaying_attack_protocol.aaa.xia
success bind to the dag

Saved HID Server = This Is The Real Host0

Enter the message = Test Replaying Attack

Data to be sent = This Is The Real Host0,Test Replaying Attack,

buf1 = This Is The Real Host0,Test Replaying Attack,

Xsock 4 new session
sent 45 bytes
```

Figure 8   Send the data using HID of host0 as a header

```
attacker@attacker-VirtualBox:~/xia-core/bin$ ./modDataExchangeClient

Stream DAG
DAG 0 -
AD:1000000000000000000000000000000000000000 1 -
HID:0000000000000000000000000000000000000000 2 -
SID:0f00000000000000000000000000000000000888

Xsock 3 created
Xsock 3 connected
Xsock 3 received 45 bytes
Received data = This Is The Real Host0,Test Replaying Attack,
```

Figure 9   The attacker captures and saves the data

```
attacker@attacker-VirtualBox:~/xia-core/bin$ ./modDataExchangeServer
Stream service started

Stream DAG
DAG 0 -
AD:1000000000000000000000000000000000000000 1 -
HID:0000000078b6604a34b211e3bb210800275a064f 2 -
SID:0f00000000000000000000000000000000000888

registered name = www_s.replaying_attack_protocol.aaa.xia
success bind to the dag

Data to be sent (Saved Data) This Is The Real Host0,Test Replaying Attack,

Xsock 4 new session
sent 45 bytes
```

Figure 10   The attacker sends the data in the next session

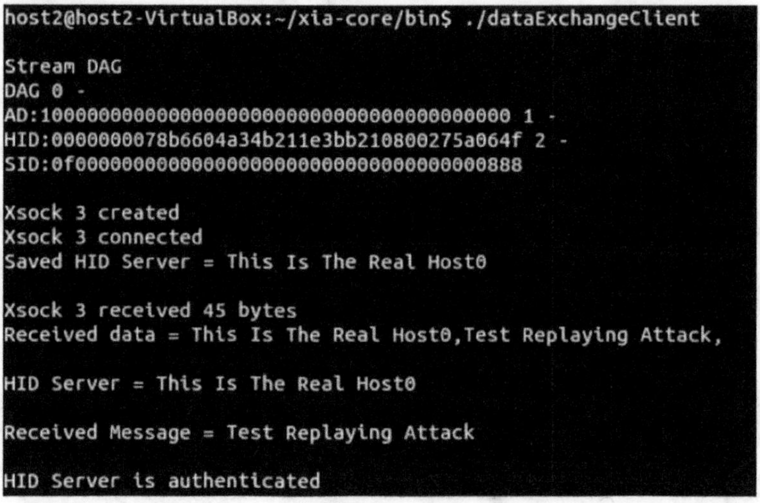

```
host2@host2-VirtualBox:~/xia-core/bin$ ./dataExchangeClient

Stream DAG
DAG 0 -
AD:1000000000000000000000000000000000000000 1 -
HID:0000000078b6604a34b211e3bb210800275a064f 2 -
SID:0f00000000000000000000000000000000000888

Xsock 3 created
Xsock 3 connected
Saved HID Server = This Is The Real Host0

Xsock 3 received 45 bytes
Received data = This Is The Real Host0,Test Replaying Attack,

HID Server = This Is The Real Host0

Received Message = Test Replaying Attack

HID Server is authenticated
```

Figure 11 Host1Receives the data from the attacker

It can be seen in Figure 10 and Figure 11 that the Attacker managed to perform the replaying attack. The Attacker replayed the data from Host0 to Host1 (as shown in Figure 10) and Host1 authenticated the Attacker as Host0 (as shown in Figure 11).

7.3 Mitigating Replaying Attack by Applying the Proposed Protocol

This case is to simulate how the proposed protocol mitigates the replaying attack. Host0 and Host1 create a session key by running the protocol. This protocol will be run in each session to create a session key that is unique in every session. The sequence diagram of this step can be seen in Figure 12.

The result from this case is, Host1 detects a replaying attack because the session key that is used by the Attacker is different than the session key that were generated by Host0 and Host1. This is because the session keys are different in every session. Once the attack is detected, Host1 terminates the session, and generates a new session key with Host0. The result of replaying attack detection by using the protocol can be seen in Figure 13.

Figure 13 represents the following processes:

- Host1 received the data that was sent by the Attacker. It is assumed that the data was captured by the Attacker in the 1st session, then he sent it to Host1 in the 2nd session.

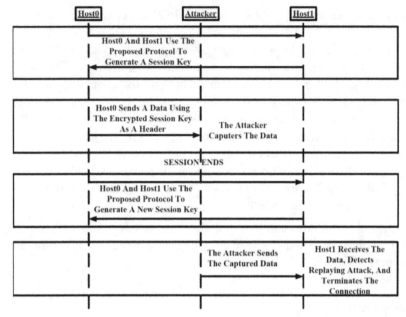

Figure 12 Sequence diagram for mitigating replaying attack scenario

```
host2@host2-VirtualBox:~/xia-core/bin$ ./modClientApps1

Stream DAG
DAG 0 -
AD:1000000000000000000000000000000000000000 1 -
HID:000000006728def3452e11e3829408002752064f 2 -
SID:0f00000000000000000000000000000000000888

Xsock 3 created
Xsock 3 connected
Xsock 3 received 418 bytes
Received data = 10385,3125,10273,12126,3125,12126,22361,23109,22361,15743,10385,
10385,17824,6752,23109,694,15842,15842,15917,12126,10385,15743,15842,12126,3125,
15743,15917,694,6752,6752,10385,3125,15842,10273,17824,15917,10273,20956,8136,69
4,10385,12126,3125,8136,20956,5922,24074,22361,24074,10273,10385,24074,694,20956
,6752,8136,3125,15917,6752,20956,10385,23109,20956,6752,15743,6752,10273,10385,2
4074,10385,test applying protocol,

Message = test applying protocol

Current Session Key = 0762917e240205e2a13104e90801a901aacd4b5c0b5370474c0035d09e
5c84910c92bc

Decryption process ...

Received Session Key = 041747e5e200b35c88a7828742ac330481ba196c07469fded10dc9364
a3905932310d0

Session Key is not valid!!!
```

Figure 13 Replaying attack detection

- In order to detect the replaying attack, Host1 separated the encrypted session key and the message.
- Host1 authenticated the session key by comparing the value of "Current Session Key" and the value of "Received Session Key". The value of "Current Session Key" is obtained when Host1 run the protocol application in 2nd session, meanwhile, the value of "Received Session Key" is obtained by decrypting the session key that was sent by Host0.
- When the comparison failed, Host1 sent out an error message. The error message was to inform that the session key is not valid (Host1 detected the replaying attack). Furthermore, Host1 terminated the session because the data that was received by Host1 was encrypted by using different session key than the current session key.

From the above figures can be concluded that the new protocol is able to detect replaying attack. In the next section the evaluation of the new protocol is presented.

8 Evaluation

This Section presents the evaluation of the proposed protocol by comparing it with the existing solutions as shown in Figure 14. The evaluation can also be seen in [46].

It can be seen in Figure 14 that:

1. The proposed protocol generates a session key with a length of 280 bits. The session key is never exchanged between hosts, therefore, the Attacker needs to guess the session key if he wants to carry out an attack (e.g., replaying or modification attacks). The possibility for the attacker to guess the session key is 2^{280}.
2. It has three messages to be exchanged while running the protocol. This amount of message is smaller than the other solutions that have four (Diffie-Hellman, Keung-Siu Protocol, and Yoon-Jeon Protocol) or five messages (Lamport's Password Authentication, Luo-Shieh-Shen Authentication Protocol, and Tseng-Jou Protocol).
3. It has four random numbers. The random numbers are used to generate the session key, a possibility for an attacker to guess four random number is $10^{70} \times 4$ (given 10 possibilities in one digit) and it is larger than a possibility to guess two random numbers (used in Diffie-Hellman and Yoon-Jeon Protocol), which is $10^{70} \times 2$. Furthermore, it is less complex than the solution that has six random numbers (Tseng-Jou Protocol).

	Year Founded	Number of Messages	Amount of Random Numbers	Number of Data Encryption	Number of Data Decryption	Key Length
Diffie-Hellman	1976	4	2	0	0	
Lamport's Password	1981	5	0	0	0	
S/KEY One-Time Password	1994	2	0	0	0	64 bits
Keung-Siu Protocol	1995	4 (Client-Server), 7 (Peer-to-peer)	4	4 (Client-Server), 8 (Peer-to-peer)	4 (Client-Server), 8 (Peer-to-peer)	
Message Binding	1997	0	0	0	0	
Timestamp	1991	0	0	0	0	
Luo-Shieh-Shen Protocol	2006	5	4	8	8	
Yoon-Jeon Protocol	2010	4	2	2	2	
Tseng-Jou Protocol	2011	5	6	4	4	
Proposed Protocol	2013	3	4	2	2	280 bits

Figure 14   Comparison of the proposed protocol with existing solutions

In addition, none of these random numbers are useless like in Yoon-Jeon and Tseng-Jou Protocols.

4. It has two times data encryption and decryption, therefore, it reduces the complexity than the other solutions that have four (Tseng-Jou Protocol and Keung-Seu Protocol between client and server) or even eight times data encryption-decryption (Keung-Seu Protocol between two clients and Luo-Shieh-Shen Authentication Protocol).

9 Conclusion and Future Work

The eXpressive Internet Architecture (XIA) is an open-source Content-Centric Network (CCN) which has potential to be standardized in future as CCN is claimed by the Future Content Networks (FCN) Group to be the Future Internet (FI). However, XIA lacks mechanism to mitigate replaying attack. Therefore, a solution for replaying attack has been proposed and implemented in this article. Nine existing solutions such as Diffie-Hellmann, Lamport's Password Authentication, S/Key One Time Password, Keung-Siu Protocol, Message

Binding, Timestamp, Luo-Shieh-Shien Authentication Protocol, Yoon-Jeon Protocol, and Tseng-Jou Protocol have been analyzed to derive the requirements for the proposed protocol. Based on the derived requirements, the solution has been developed.

The protocol has been implemented in XIA prototype and has been proven to be able to mitigate the replaying attack. The proposed protocol has the following properties: First, There is a unique session key for each host in every session. Second, there is a checking process to ensure that the session key that is generated at each host is the same. Third, it has mechanisms to encrypt the messages and to conceal the identity of the hosts.

The proposed protocol has been evaluated to have more advantages over the reviewed existing solutions. It is more secure by having session key with length of 280 bits. Moreover, it is less complex as none of the random numbers used in the protocol are worthless. By applying the proposed protocol, XIA is now able to mitigate all of the reviewed attacks.

According to the current standard [45], a session key with a length of up to 280 bits is secure. In the future, when 280 bits is not enough, the size of the session key can be extended.

10 Standardization Candidate

The standardization of the future network architectures is a responsibility of the ITU-T Study Group 13 (SG 13) [47]. The SG 13 consists of several groups, and Focus Group Future Networks (FG FN) is one of them, and its aim is to collect FI architectures and technologies to be standardized [48]. The new protocol that is presented in this paper can be standardized to be used in XIA, as it causes XIA to be robust against all of the reviewed attacks. Furthermore, the protocol is more secure and more efficient than others replaying attack solution, thus, the protocol can also be standardized for the current Internet network.

References

[1] Anja Feldmann, "Internet Clean-Slate Design: What and Why?," in *SIGCOMM Computer Communication Review*. 2007, pp. 59–64. Volume 37, Number 3, ACM.

[2] Ashok Anand, Fahad Dogar, Dongsu Han, Boyan Li, Hyeontaek Lim, Michel Machado, Wenfei Wu, Aditya Akella, David Andersen, John

Byers, Srinivasan Seshan, and Peter Steenkiste, "XIA: An Architecture for an Evolvable and Trustworthy Internet," in *Proceedings of the tenth ACM Workshop on Hot Topics in Networks (HotNets-X)*. 2011, pp. 1–32. Article No. 2, ACM.

[3] John Day, Ibrahim Matta, and Karim Mattar, "Networking is IPC: A Guiding Principle to a Better Internet," in *Proceedings of the 2008 ACM CoNEXT Conference. 2008*, pp. 1–6. Article Number 67, ACM.

[4] Bernd Reuther and Paul Müller, "Future Internet Architecture - A Service Oriented Approach," in *In IT - Information Technology*, Volume 50, Number 6, 2008, pp. 1–7.

[5] Denis Martin, Lars Völker, and Martina Zitterbart, "A flexible framework for Future Internet design, assessment, and operation," *Journal Computer Networks: The International Journal of Computer and Telecommunications Networking*, pp. 910–918. Volume 55 Issue 4, March 2011.

[6] Ivan Seskar, Kiran Nagaraja, Sam Nelson, and Dipankar Raychaudhuri, "MobilityFirst Future Internet Architecture Project," in *Proceeding of: AINTEC '11, Asian Internet Engineering Conference*, 2011, pp. 1–3.

[7] Robert Broberg, Matthew Caesar, Douglas Comer, Chase Cotton, Michael J. Freedman, Andreas Haeberlen, Zachary G. Ives, Arvind Krishnamurthy, William Lehr, Boon Thau Loo, David Mazires, Antonio Nicolosi, Jonathan M. Smith, Ion Stoica, Robbert van Renesse, Michael Walfish, Hakim Weatherspoon,and Christopher S. Yoo, "The NEB-ULA Future Internet Architecture," *Lecture Notes in Computer Science*, pp. 1–24. Volume 7858, 2013.

[8] Lixia Zhang, Deborah Estrin, Jeffrey Burke, Van Jacobson, James D. Thornton, Diana K. Smetters, Beichuan Zhang, Gene Tsudik, KC Claffy, Dmitri Krioukov, Dan Massey, Christos Papadopoulos, Tarek Abdelzaher, Lan Wang, Patrick Crowley, and Edmund Yeh, "Named Data Networking (NDN) Project," pp. 1–26, 2010.

[9] Rowan Klöti, "OpenFlow: A Security Analysis," M.S. thesis, Swiss Federal Institute of Technology Zurich, 2013.

[10] Jie Wang, *Computer Network Security: Theory and Practice*, Higher Education Press, 2009.

[11] Ltd Hangzhou H3C Technologies Co., "Attack Prevention Technology White Paper," 2008.

[12] Emmett Dulaney, *CompTIA Security+ Study Guide*, Wiley, Indianapolis, 4th edition, 2009

[13] Jelena Mirkovic, Sven Dietrich, David Dittrich, and Peter Reiher, *Internet Denial of Service: Attack and Defence Mechanisms*, Prentice Hall, 2005

[14] Mark Ciampa, *Security Plus Guide to Network Security Fundamentals*, Cengage Learning, 3rd edition, 2009.

[15] Ling Dong and Kefei Chen, *Cryptographic Protocol: Security Analysis Based on Trusted Freshness*, Springer, 2012.

[16] Hamid Jahankhani, David Lilburn Watson, Gianluigi Me, and Frank Leonhardt, *Handbook of Electronic Security and Digital Forensics*, 2010

[17] Hannes Gredler and Walter Goralski, *The Complete IS-IS Routing Protocol*, Springer, 2004.

[18] Future Internet Assembly (FIA) Future Content Networks (FCN) Group, "Technical Report. Why do we need a Content Centric Future Internet?," pp. 1–23, 2009.

[19] M. Rahamatullah Khondoker, Abbas Siddiqui, Bernd Reuther, and Paul Müller, "Service Orientation Paradigm in Future Network Architectures," in *Sixth International Conference on Innovative Mobile and Internet Services in Ubiquitous Computing (IMIS-2012)*, 2012, pp. 346–351.

[20] Paul Müller, Bernd Reuther, and Markus Hillenbrand, "Future Internet: A Service-Oriented Approach - SONATE," in *Würzburg Workshop on Visions of Future Generation Networks (EuroView2007)*, 2007, pp. 1–35.

[21] Oliver Hanka and Hans Wippel, "Secure Deployment of Application-Tailored Protocols in Future Networks," in *Proceedings of the Second International Conference on the Network of the Future (NoF 2011)*, 2011, pp. 10–14.

[22] Thomas Gamer and Hans Wippel, "A Collaborative Attack Detection and its Challenges in the Future Internet," in *Proceedings of the Joint ITG, ITC, and Euro-NF Workshop "Visions of Future Generation Networks" (EuroView)*, 2010, pp. 1–2.

[23] Hsu-Chun Hsiao, Tiffany Hyun-Jin Kim, Adrian Perrig, Akira Yamada, Samuel C. Nelson, Marco Gruteser, and Wei Meng, "LAP: Lightweight Anonymity and Privacy," in *Proceedings of the IEEE Symposium on Security and Privacy*. 2012, pp. 506–520, IEEE Computer Society.

[24] Hsu-Chun Hsiao, Tiffany Hyun-Jin Kim, Sangjae Yoo, Xin Zhang, Soo Bum Lee, Virgil Gligor, and Adrian Perrig, "STRIDE: Sanctuary Trail Refuge from Internet DDoS Entrapment," in *Proceedings of the ACM Symposium on Information, Computer and Communications Security (ASIACCS)*. 2013, pp. 415–426, ACM.

[25] Tiffany Hyun-Jin Kim, Lin-Shung Huang, Adrian Perrig, Collin Jackson, and Virgil Gligor, "Accountable Key Infrastructure (AKI): A Proposal for a Public-Key Validation Infrastructure," in *Proceedings of the 22nd*

international conference on World Wide Web. 2013, pp. 679–690, International World Wide Web Conferences Steering Committee.

[26] Eleni Trouva, Eduard Grasa, John Day, Ibrahim Matta, Lou Chitkushev, Patrick Phelan, and Miguel Ponce de Leon Steve Bunch, "Is the Internet an unfinished demo? Meet RINA!," in *TERENA Networking Conference*, 2011, pp. 1–12.

[27] Gowtham Boddapati, John Day, Ibrahim Matta, and Lou Chitkushev, "Assessing the Security of a Clean-Slate Internet Architecture," in *Proceedings of the Seventh Workshop on Secure Network Protocols (NPSec).* 2012, pp. 1–6, 20th IEEE International Conference Network Protocols (ICNP).

[28] Jeremiah Small, "Patterns in Network Security: an Analysis of Recursive Inter-Network Architecture Security Module Efficiency," M.S. thesis, Boston University, 2012.

[29] Feixiong Zhang, Kiran Nagaraja, Yanyong Zhang, and Dipankar Raychaudhuri, "Content Delivery in the MobilityFirst Future Internet Architecture," in *Sarnoff Symposium (SARNOFF)*, 35th IEEE, 2012, pp. 1–5.

[30] MobilityFirst Project Team, "MobilityFirst: A Robust and Trustworthy Mobility-Centric Architecture for the Future Internet," Tech. Rep., 2010.

[31] Paolo Gasti, Gene Tsudik, Ersin Uzun, and Lixia Zhang, "DoS and DDoS in Named-Data Networking," 2012, pp. 1–10. Volume abs/1208.0952.

[32] Jad Naous, Michael Walfish, Antonio Nicolosi, David Mazieres, Michael Miller, and Arun Seehra, "Verifying and Enforcing Network Paths With ICING," in *Proceedings of the Seventh Conference on emerging Networking EXperiments and Technologies.* 2011, pp. 1–12. Article No. 30, ACM.

[33] XIA Project Team, "XIA Prototype," https://github.com/XIA-Project/xia-core/wiki, 2013, [Online; Accessed on 01-August-2013].

[34] Whitfield Diffie and Martin E. Hellman, "New Directions in Cryptography," *Journal IEEE Transactions on Information Theory*, pp. 644–654. Volume 22 Issue 6, 1976.

[35] Leslie Lamport, "Password Authentication With Insecure Communication," *Magazine Communications of the ACM*, pp. 770–772. Volume 24 Issue 11, 1981.

[36] Neil Haller, "The S/KEY One-Time Password System," in *Proceedings of the Internet Society Symposium on Network and Distributed Systems*, 1994, pp. 151–157.

[37] Stephen Keung and Kai-Yeung Siu, "Efficient Protocols Secure Against Guessing and Replay Attacks," in *Proceedings, Fourth International Conference on Computer Communications and Networks*, 1995, pp. 105–112.

[38] Tuomas Aura, "Strategies against Replay Attacks," in *Proceedings of the 10th IEEE workshop on Computer Security Foundations CSFW'97*, 1997, pp. 59–68.

[39] Cdric Adjih, Daniele Raffo, and Paul Mhlethaler, "Attacks Against OLSR: Distributed Key Management for Security," *2nd OLSR Interop/Wksp.*, pp. 1–7, 2005.

[40] Jia-Ning Luo, Shiuhpyng Shieh, and Ji-Chiang Shen, "Secure Authentication Protocols Resistant to Guessing Attacks," *Journal of Information Science and Engineering*, pp. 1125–1143. Volume 22 No. 5, 2006.

[41] Eun-Jun Yoon and Il-Soo Jeon, "An efficient and secure Diffie Hellman key agreement protocol based on Chebyshev chaotic map," *Communications in Nonlinear Science and Numerical Simulation*, pp. 23832389. Volume 16, Issue 6, 2010.

[42] Huei-Ru Tseng and Emery Jou, "An Efficient Anonymous Key Agreement Protocol Based on Chaotic Maps," in *IEEE 13th International Conference on High Performance Computing and Communications (HPCC)*, 2011, pp. 752–757.

[43] Tsuji Takasuke, "A One-Time Password Authentication Method," M.S. thesis, Graduate School of Engineering, Kochi University of Technology, 2002.

[44] Sung-Ming Yen and Kuo-Hong Liao, "Shared authentication token secure against replay and weak key attacks," in *Information Processing Letters*. 1997, pp. 77–80. Volume 62 Issue 2, Elsevier North-Holland, Inc.

[45] Elaine Barker, William Barker, William Burr, William Polk, and Miles Smid, "Recommendation for Key Management Part 1: General (Revision 3)," pp. 1–147, 2012.

[46] Beny Nugraha, Rahamatullah Khondoker, Ronald Marx, and Kpatcha Bayarou, "A Mutual Key Agreement Protocol To Mitigate Replaying Attack In eXpressive Internet Architecture (XIA)", in *ITU Caleidoscope Academic Conference*, pp. 233–240. 2014.

[47] "ITU-T SG13: Future networks including cloud computing, mobile and next-generation networks," http://www.itu.int/en/ITU-T/studygroups/2013-2016/13/Pages/default.aspx, Online; accessed 06-Dec-2013.

[48] "ITU-T FG FN: Focus Group on Future Networks (FG FN)," http://www. itu.int/en/ITUT/focusgroups/fn/Pages/Default.aspx, Online; accessed 06-Dec-2013.

Biographies

Beny Nugraha received his dual masters degree – International Master Degree Program from Bandung Institute of Technology (Indonesia) and Hochschule Darsmtadt (Germany) in 2013. In order to finish his Master Degree in Germany, he received a scholarship from the Indonesian Directorate General of Higher Education. Currently, he is a lecturer at the department of Electrical Engineering in Mercu Buana University located in Jakarta, Indonesia. His research is mainly about network security, currently he is focusing on the security of Future Internet Architectures and cloud computing.

Since 2010, **Rahamatullah Khondoker** has been working towards his PhD degree on "Description and Selection of Communication Services for Service Oriented Network Architectures (SONATE)" at the University of

Kaiserslautern in Germany. He was awarded from Ericsson, Germany in the year 2008 and from the FIA Research Roadmap group in October 2011. Currently, he is affiliated with the Fraunhofer SIT located in Darmstadt, Germany. He worked with the DFG project (PoSSuM), BMBF projects (G-Lab, G-Lab DEEP, Future-IN), and EU projects (PROMISE, EuroNF). Currently, he is focusing on the security of Future Internet Architectures, Software-Defined Networking (SDN), and Network Function Virtualization (NFV).

Ronald Marx is the deputy head of the "Mobile Networks" at the Fraunhofer Institute for Secure Information Technology (SIT). He received his diploma in computer science at the Technical University of Darmstadt (TUD). Since 2005, he was involved in numerous projects, as project staff and project manager. His work focuses on the security aspects in next generation networks (NGN), the mobility and identity management and voice over IP communications.

DR. KPATCHA BAYAROU received his Diploma in electrical engineering/ automation engineering in 1989, a Diploma in computer science in 1997, and his Doctoral degree in computer science in 2001, all from the University of

Bremen in Germany. He joined the Fraunhofer Institute for Secure Information Technology (Fraunhofer SIT) in 2001. He is the head of the "Mobile Networks" department that focuses on Cyber Physical Systems and Future Internet including vehicular communication. Dr. Bayarou managed several EU and nationally funded projects and published several conference papers related to security engineering of mobile communication systems, mobile network technology, and NGN (Next Generation Networks).

Challenges of Security Assurance Standardization in ICT

Marcus Wong

Huawei Technologies (USA), Bridgewater, New Jersey, USA, mwong@huawei.com

Received: September 5, 2014;
Accepted: November 10, 2014

Abstract

The explosion of mobile broadband growth has created a greater demand on the operators and vendors working together to place more and more telecom gears into wireless networks at a record pace to satisfy the users' insatiable appetite for mobile data. The desire for undiminished security coupled with more sophisticated attacks in an ICT world where the traditionally closed telecom networks are going through a change of open architecture, open platform, and virtualization, the entire telecommunication community has taken a proactive approach to re-evaluate the security assurance process to ensure that the products are as secure as ever. The operators and the vendors have come together under the roof of 3GPP to create such a security assurance standards to be applied, recognized, and accepted in all areas for which 3GPP network products are sold and marketed. This paper will examine the many issues, hurdles, and challenges of the standardization of security assurance.

Keywords: Security Assurance, 3GPP, standards.

1 Introduction

The market needs and lacking of a security assurance for telecommunication sector coupled with the explosion of mobile broadband growth have made the condition ripe to start putting the focus on an industry-wide security

Journal of ICT, Vol. 2, 187–200.
doi: 10.13052/jicts2245-800X.226

assurance process for telecommunication products. Attacks have become more and more sophisticated; attackers have become more and more intelligent; and the attack tools have become more and more advanced. At the same time, the wireless networks have become more and more open in terms of architecture and visibility. The combination of these events has called upon the entire telecommunication community to re-evaluate the approach and the process to ensure that the products are more and more secure. This leads to an international security assurance standards for the wireless telecom product so that one process can be applied to every market, every product and for every stakeholder. As a result, expert members in the Third Generation Partnership Program (3GPP) security group have committed themselves to create such a security assurance process.

2 The Challenges

Riding the wave of 3G, 4G and the forthcoming 5G, operators have introduced a plethora of new services that not only rely on new products and features to be developed quicker than ever before but also that these products and features are rolled out to market place at a record pace. The users and the operators still demand and expect the same undiminished security values they have become accustomed to and offered by the vendors as the network security landscape and threat models are constantly evolving. The challenges are real. The interests are high for customers, governments, and vendors alike to ensure that the telecommunication products and the networks are more secure than ever.

2.1 Security Challenges

Today's telecommunication networks have become more open, and at the same time more sophisticated and more intelligent. We are relying on the communication networks and connectivity more than ever. Information, tools, and knowledge about networks and network security are readily available to anyone who has the desire and determination to learn about anything or gain a great deal from it, including those who attempt to seek financial gains or those who seek to create damage and disruption. When the information and knowledge are in the wrong hands, along with the more powerful machines and tools of today (e.g. PCs, tablets, smart phones, etc.), it is becoming increasingly evident that malicious misuse of the learned or gained knowledge can lead to serious disruption of communication networks and network services. Being able to communicate anytime and anywhere also means that attackers are able to launch attacks anytime and anywhere. The potential losses can be great in terms of productivity losses, financial losses and information losses. It has become vitally important to keep the communication systems and network more secure than ever. Ensuring the security of systems and networks has become one of the toughest challenges for the entire telecom ecosystem in the foreseeable future. Some of these challenges include:

- Identity management
- Virus, malware and botnets
- Internet-based attacks
- Industrial espionage and sabotage
- Privacy laws and regulations
- Awareness, education, and training

These challenges also bring along an assortment of many potential threats and risks to the products, networks, and services:

Figure 1 Threat and attack model

- Physical tampering
- Denial of service and attacks on the networks
- Compromise of authentication credentials
- Man-in-the-middle attacks on the networks
- Intercepting and modification of user's data
- Rogue network equipment
- Mis-configuration
- Radio resource tampering
- Hackers

2.2 Market Place Challenges

In addition to building secure products, meeting these challenges also requires that a global security assurance process to be developed so that the security of the products can be demonstrated in a systematic approach. The needs for such security assurance are also driven by the market for such a transparent process, for instances in the Indian market and other markets around many parts of the world. Making claims that a particular product is either secure or insecure without substantiating the claims is simply irresponsible. From the operator's perspective, they demand that the products they place in their networks are secure and are developed with the highest integrity as they not only have their reputation to protect, the user's security and privacy to protect, but also the laws and regulations to comply with as there are numerous such regional and national laws and regulations regarding the protection of user data and user privacy. From the vendor's perspective, they want to ensure that not only the operator's requirements are met and that their products are secure, but also that they can keep up with the demand for faster feature and product development without compromise in security as they too, have a reputation to protect. At the same time, the vendors also want an environment to ensure that their products can be used in all markets of the globe without the need to customize and certify the products for each market. To reach this goal, the operators and the vendors have come together in the telecommunication industry create such a security assurance standards to be applied, recognized, and accepted in all areas for which network products are sold and marketed. Bottom line is that the telecommunication industry needs such a security assurance process so that every stakeholder can benefit from it. Products built to the security assurance specification and having gone through the security assurance process will be able to withstand any unsubstantiated claims about the security of the products.

2.3 Regulatory Challenges

Regulators have spoken loudly and clearly with laws regulations put in place in various regions around the world covering data security, user privacy, and with strict requirements for telecommunication products when the systems and networks provide vital services to serve the government and regulator communities in addition to serve the general public. Some examples of these regulations and directives include EU's Data Protection Directive, UK's Data Protection Act, Canada's Privacy Act, Japan's Personal Information Privacy, China's Provisions for Administration of Information Service, so on and so forth. The list goes on and on. It should be noted that the varying regulators should not overly burden the industry with unnecessary and impractical rules and regulations that are difficult to harmonize and thereby creating a barrier to inhibit innovation.

3 Requirements of Security Assurance

Meeting the challenges head-on requires that the security assurance process must be transparent, collaborative, global, standards-based, and practical to be effective. Requirements from operators, vendors, and regulators need to be taken into account fully to build up the security assurance process. Market needs drive products and services, which in turn drive requirements on the vendors, operators and regulators. The market place is unyieldingly unwilling to compromise even in the face of increased threats and risks presented. Regulatory requirements are also an important and necessary component of the security assurance process. All and all, it requires the best of vendors, operators, and regulators to come together to define ways of ensuring the security of products, systems, networks, and the users. The network operators also have requirements and demand the best of vendors and suppliers not only to build the best and most secure products, but also to provide indisputable evidence, in the form of assurance and certification. The operators have moral, legal, and financial requirements and obligations to ensure the security and privacy of its customers – the very users who contribute to the growth and success of the operators. Besides building the most secure product, vendors and suppliers are also required to ensure that not only they follow the most strict industry security best practices, but also receive the necessary accreditation to ensure that they are held to the highest standards.

4 Security Assurance Process

The security assurance process is certainly not new. The process already exists for many reasons for IT products. The process can be used for networking products to a certain extent but with some degree of limitations. The most notable of these developments is the Common Criteria (CC) and the Common Criteria Recognition Arrangement (CCRA) which combined three standards: Information Technology Security Evaluation Criteria (ITSEC), Canadian Trusted Computer Product Evaluation Criteria (CTCPEC), and Trusted Computer Security Evaluation Criteria (TCSEC) originally developed by the governments of European countries, Canada, and US respectively.

4.1 Goals of Security Assurance

In recent years, some of the legal requirements have put the vendors and operators in a dilemma as they risk the possibility of financial losses in terms of fines originating from government enforced penalties and lost business opportunities as result of negative publicity when these requirements are not met. Vendors have to show not only that they have followed the strict operator requirements and legal requirements, but also industry best practices that they have built the products to the highest degree of standards, but also with highest degree of security and integrity. Furthermore, vendors may have to repeat the process in every market and region where their products are deployed.

There should be a globally standardized process in demonstrating the security of products, systems, and networks. This process has to be systematic that every stakeholder can work with and rely upon. The stakeholders include vendors, suppliers, network operators, regulators, government agencies, etc. With so much at stake, it is easy to understand the goal of this approach – to specify the network product security assurance requirements that are necessary to protect against unwanted access to the product, its operating systems, and running applications. The security assurance requirements to be developed and specified should be based on threat and attacker models that are applicable to the functions the products are designed to perform, including generic IT and communication functions. The security assurance requirements are of course in addition to any basic functional requirements and feature requirements of the products to be developed. For instance, a base station will be developed with set of required core functions (e.g. RF, communication, etc.) with the security assurance requirements in mind as these security assurance requirements are taken as baseline for building products that are not only functional, but also demonstrably secure. This systemic approach becomes the "Security Assurance Process".

4.2 Challenges of Security Assurance

Although CC and CCRA have existed for many years and have gained international acceptance with more than twenty member countries around the world, but the framework and infrastructure were developed mainly to focus on IT products as well as computer products, and were originally developed to serve the government and intelligence markets.

Many attempts to apply the CC and the CCRA framework for certain telecommunication products have shown the process to be both intensive and time consuming and may not meet the need for products to reach market timely with many of the features and products that are required to offer value added services to the users. Obtaining CC certification for products, even at particular levels acceptable to private communication networks (e.g. EAL 3) would mean thousands of dollars and many more man-hours spent with one of the CC-accredited laboratories around the world before certification can be obtained. Obtaining certification for higher level of assurance requires even greater efforts, more time and more money. To that effect, CC has been around many years and has served a good purpose. Many members within the telecommunication community felt that it does not address the constant changing needs and requirements of the telecommunication industry since it was designed for a different class of products and it may prove difficult to adopt it to accommodate the security assurance requirements of telecommunication products without substantial modifications. Trying to fit the networking products into the CC framework for the purpose of security assurance and accreditation has proven to be awkward, time-consuming, and expensive even though CC is not without its merits. The lessons learned and experience gained through CC and the CCRA framework will serve as a solid basis for developing security assurance process in the telecommunication environment.

Another challenge is having the stakeholders to endorse the process once it is completed. For CC and CCRA, it took quite some time for it to be recognized in twenty-plus countries, mostly through governmental efforts. Without these efforts and relying on the industry leverage alone may prove difficult even though 3GPP is an internationally recongnized standards body which produces de facto specification for wireless systems around the world. Recognition of

3GPP standards is often quite different than recognition of security assurance standards such as interoperability aspects.

Yet another challenge is choosing the right threat model and security framework in the process. There are various threat models and security framework created for different purposes, such as STIX, short for Structured Threat Information Expression, STRIDE, short for Spoofing, Tampering, Reputation, Information Disclosure, Denial of Service, Elevation of Privilege, or ITU's X.805 Security Architecture for systems providing end-to-end communications. Though there are similarities among them, but like CC, they were developed for specific cases making adaptation in telecommunication difficult. There is no right or wrong threat model and everyone has its own merits.

4.3 3GPP Security Assurance

A good example of putting the lessons learned through CC and the CCRA framework to use is shown by 3GPP in its Security Assurance Methodology (SECAM) and Security Assurance Specifications (SCAS) activities. This work is also done in conjunction with GSMA's Network Equipment Security Assurance Group (NESAG) where as 3GPP defines the security baseline specifications including test cases for evaluating the results while NESAG defines the framework for accrediting evaluation laboratories (including vendor evaluation laboratories and third-party evaluation laboratories) and resolution in case of disputes between the vendors and operators.

SECAM and SCAS are seen as a positive development of such a security assurance methodology specifically for the 3GPP products, as the first attempt to evolve into an international standards purely from the industry perspective. SECAM and SCAS are intended to be a comprehensive process for which all network product and network product when the process is fully implemented. It starts with identification of the threats and risks associated with each product. Although there may be many functions within each product (e.g. encryption, authentication, etc.), the focus is to perform the threat and risk analysis on the entire product as a whole so that the security requirements along with security assurance specification can be developed. Next, the security requirements will be developed for that product, which may be done in modular fashion, for example based on functional components, to afford the flexibility of applying these modules of requirements to different products with same or similar functions without duplicating efforts needed to develop security assurance requirements for the same function in another product.

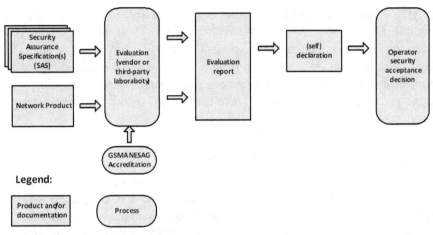

Figure 2 3GPP Security Assurance Process

The security requirements take into account threat model and risk analysis, attacker potentials and capabilities, environmental variations, etc. to resist all known attacks (both current and anticipated). A testing process (e.g. evaluation by an accredited laboratory that can be either a vendor or a third-party independent laboratory) then follows with test cases and testing methodology that will be able to produce verifiable and repeatable results from the security testing of the product. All product will be tested vigorously and comprehensively to ensure that that the product being built and tested will conform to the security requirements. This is the same philosophy of assuming nothing, believing in no one, but checking everything with a multi-layered "many hands" and "many eyes" approach to independent verification in order to reduce the risk of insecure products being distributed.

The final step in the security assurance process is for the operator to gain confidence after the product has clearly demonstrated meeting all of the defined security requirements and passing all of the verifications. Acceptance is also backed by rigorous and yet robust audit mechanism where verifiability, traceability, and disputes can be resolved.

More than ever, the operators demand greater uniformity in terms of requiring the same security assurance from all vendors, especially in a multi-vendor deployment environment. The users demand unequivocal guarantee in terms of security and privacy. And finally the governments have placed stringent requirements on the service providers to deploy secure network equipments and networks in the name of national security as more and more governments are relying less on dedicated networks and more on private networks to carry their traffic. Greater emphases are placed on the security assurance of telecom products, as the entire industry has experienced an evolution of migrating to open architecture and open platform from a traditionally considered closed environment. This, in turn will drive the telecommunication community to create the standards necessary to achieve the desired security assurance for the products developed to comply with specifications.

4.4 3GPP Security Assurance Approach

In order for the security assurance process to work, a multi-layered approach needs to be taken where the security assurance process is to be developed as an open and transparent process where all stakeholders with their vast expertise and experience from aspects of telecommunication, information technology, networking, and security contribute toward a common goal. The ultimate goal is not only to give the operators assurance that the products built are secure, but also to give assurance to all stakeholders that the products built are secure against the known attacks at the time of deployment.

Every stakeholder across all regions also needs to sign on to and agree to the process. Working with other standardization bodies, such as ETSI, IETF, ITU, 3GPP, etc and with regulators will ensure mutual acceptance once a particular product has successfully gone through the security assurance process is not alone in perfecting the security assurance process. Other organizations' expertise is also very helpful as those have gone through similar efforts such as National Institute of Standards and Technology (NIST) Security Content Automation Program (SCAP), the Security Automation and Continuous Monitoring (SACM) in IETF, and SECAM and SCAS in 3GPP even though sometimes the efforts by disjoint organizations would appear in random. Mutual recognition and mutual acceptance equate to a single and fair process for all and goes a long way to ensure the success of the entire process.

Though the aim is for the operators to gain confidence about the security of the products, but since the process is open and every step within the process can be documented and substantiated, it should make it very easy for all stakeholders, whether they are vendors, system operators, regulators or the like to realize the transparency within the process and to accept the products with a great deal of confidence that the products are secure, once the rigorous security assurance process has been followed. It is important to emphasize once again that such a security assurance must be transparent, collaborative, global, standards-based and practical to be effective. It is noted that as a

Figure 3 International Security Assurance Collaboration

standards-developing organization (SDO), 3GPP has made great strides to achieving the goal of security assurance through the study items and work items that have produced various technical reports such as 3GPP Study on Security Assurance Methodology for 3GPP Network Products Technical Report, 3GPP Pilot development of Security Assurance Specification (SCAS) for MME network product class Technical Report, and 3GPP Security Assurance Methodology for 3GPP network products Technical Report. Once the work is done, 3GPP will have produced security assurance specification for all 3GPP products, starting with the Mobility Management Entity and extending to other network product class such as evolved NodeBs (eNB), Serving Gateways (SGW), etc.

In summary, here is the approach and process taken in 3GPP to develop a security assurance specification:

1. Establish and methodology for 3GPP security assurance
2. Create testable requirements and test cases
3. Develop specification on security assurance for a particular 3GPP product (e.g. MME)
4. Test the process
5. Extend the process for all all 3GPP products.

4.5 Security Assurance Around the Globe

Recognizing the importance of security assurance in ICT, many other SDOs have accelerated their pace in bringing more awareness through the development and work on security assurances. Work in CC is continuing in terms of making it more "user-friendly" by considering the different levels of assurance. IETF has taken on the work of creating use cases for Endpoint Security Posture Assessment and has gone through several iterations of internet drafts. GISFI is working closely with 3GPP in addressing the security assurance requirements originating from the Indian telecom market. CESG, though not a SDO, has also development a commercial product scheme or CPA for short for the UK market to provide assured commercial security products for users who have a need for information assurance. With network virtualization and SDN on everyone's mind, the European Union Agency for Network and Information Security has also developed a set of assurance criteria to assess the risks of adopting cloud services, obtain assurance from the cloud providers, and reduce the assurance burden on the cloud providers. Similar activites in ITU produced Entity Authentication Assurance Framework in ITU-T's X.1254. The list goes on and on, but none bigger than the challenges taken up in 3GPP.

5 Conclusion

Now is the time to step up and address future global security challenges today. The team of security experts representing the operators, vendors, and regulators in 3GPP have come together in unison to focus on the security assurance process by leveraging the expertise and knowledge gained from years of creating standards. Making claims one way or another about the security of the products, systems, or networks based on misconstrued or misinformed belief without any concrete evidence is not only irresponsible but they are also not valid reasons to accept or reject a particular product. The network of tomorrow starts today. Getting away from the cycles of break-and-fix to drive the security assurance message home will ensure the success of verified and certifiable security claims. Security begins with a commitment coupled with a solid foundation of understanding the threats, defining a security assurance process, and going through vigorous testing leading to verification. It is also important for the market that a successful verification as a result of the security assurance process should be internationally recognized and accepted as the process itself is an open process. Following the security assurance process will not be beneficial for the telecommunication industry, but a win-win proposition for the entire ecosystem.

References

[1] 3GPP TR 33.805, Study on Security Assurance Methodology for 3GPP Network Products

[2] 3GPP TR 33.806, Pilot development of Security Assurance Specification (SCAS) for MME network product class

[3] 3GPP TR 33.916, Security Assurance Methodology for 3GPP network products

[4] Canada's Privacy Act, http://www.priv.gc.ca/leg_c/leg_c_a_e.asp

[5] Common Criteria for Information Technology Security Evaluation, Version 3.1 Release 4, September 2012

[6] The CC and CEM documents: http://www.commoncriteriaportal.org/cc/

[7] The CCRA introduction: http://www.commoncriteriaportal.org/ccra

[8] CCRA Licensed Laboratories: http://www.commoncriteriaportal.org/labs/

[9] EU Directive 95/46/EC, The Data Protection Directive

[10] CESG Commercial Product Assurance (CPA) Scheme: http://www.cesg.gov.uk/servicecatalogue/Product-Assurance/CPA/Pages/CPA.aspx

[11] IETF Internet Draft: "Endpoint Security Posture Assessment – Enterprise Use Cases"

[12] Cloud Computing Information Assurance Framework: http://www.enisa.europa.eu

[13] ITU-T X.1254: Entity Authentication Assurance Framework

Biography

Marcus Wong Wireless Security Research and Standardization, Huawei Technologies (USA).

Marcus received his Master of Arts Degree in Computer Science from Queens College of City University of New York (USA). He has over 20 years of experience in the wireless network security field with AT&T Bell Laboratories, AT&T Laboratories, Lucent Technologies, and Samsung's Advanced Institute of Technology. He holds Certification of Information System Security Professional (CISSP) from the prestigious International Information Systems Security Certification Consortium (ISC2).

Marcus has concentrated his research and work in many aspects of security in wireless communication systems, including 2G/3G/4G mobile networks, Personal Area Networks, and satellite communication systems. Marcus joined Huawei Technologies (USA) in 2007 and continued his focus on research and standardization in 3GPP, WiMAX Forum, IEEE, and IETF security areas. As an active contributor in the Wireless World Research Forum (WWRF), he has shared his security research on a variety of projects contributing toward whitepapers, book chapters, and speaking engagements.

In the past, Marcus has held elected official positions in both WWRF and 3GPP, serving as the vice-Chairman of WWRF Working Group 7 (Security and Trust working group) from 2007 to 2012 and as the vice-Chairman of 3GPP SA3 (Service & System Aspect, Security Group) from 2009 to 2011 respectively. He also served as guest editor in the IEEE Vehicular Technology magazine. He also has published a number of journal papers and whitepapers in leading publications, including that of the Journal of Cyber Security and Mobility. In addition, he has numerous patents granted and/or pending.